Stable and Efficient Cubature-based Filtering in Dynamical Systems

Dominik Ballreich

Stable and Efficient Cubature-based Filtering in Dynamical Systems

Dominik Ballreich
University of Hagen
Hagen, Germany

Dissertation – University of Hagen, 2017

ISBN 978-3-319-62129-6   ISBN 978-3-319-62130-2   (eBook)
DOI 10.1007/978-3-319-62130-2

Library of Congress Control Number: 2017947687

© Springer International Publishing AG 2017
This work is subject to copyright. All rights are reserved by the Publisher, whether the whole or part of the material is concerned, specifically the rights of translation, reprinting, reuse of illustrations, recitation, broadcasting, reproduction on microfilms or in any other physical way, and transmission or information storage and retrieval, electronic adaptation, computer software, or by similar or dissimilar methodology now known or hereafter developed.
The use of general descriptive names, registered names, trademarks, service marks, etc. in this publication does not imply, even in the absence of a specific statement, that such names are exempt from the relevant protective laws and regulations and therefore free for general use.
The publisher, the authors and the editors are safe to assume that the advice and information in this book are believed to be true and accurate at the date of publication. Neither the publisher nor the authors or the editors give a warranty, express or implied, with respect to the material contained herein or for any errors or omissions that may have been made. The publisher remains neutral with regard to jurisdictional claims in published maps and institutional affiliations.

Printed on acid-free paper

This Springer imprint is published by Springer Nature
The registered company is Springer International Publishing AG
The registered company address is: Gewerbestrasse 11, 6330 Cham, Switzerland

For my parents

Foreword

The work of Dominik Ballreich, entitled *Stable and Efficient Cubature-Based Filtering in Dynamical Systems*, treats the topic of deterministic numerical integration in the context of optimal filtering. The latter algorithms play a key role in optimal control and signal processing but also in economics and the social sciences. The task of prediction and estimation of dynamical systems with noisy measurements and unobserved variables can be solved using the Kalman filter recursions, which are a sequence of conditional expectation values of the system functions, given the data. These expectations can be computed using several methods, including numerical integration (cubature). In the last years, this method has been becoming popular in the engineering literature, starting with the unscented Kalman filter (UKF). Later, Gauss-Hermite integration revealed a well-known problem, the blowing up of integration points. Actually, the computational burden grows exponentially with the dimension of the state space. Here, the sparse grid methods of Smolyak led to a breakthrough. However, the stability of the method must be improved, especially for applications in recursive equations.

At this point, Ballreich shows how to construct nearly exact solutions of the moment equations, where the given degrees of freedom are used to achieve high stability, connected with positive weights. Moreover, it is shown how the cubature rules of Arasaratnam and Haykin and the unscented transform can be derived, giving the latter a theoretical foundation. The work also discusses several nonlinear benchmark problems, such as the coordinated turn model, the Kitagawa and Ginzburg-Landau model, or the chaotic Lorenz model. Here, the efficient and stable integration algorithms lead to superior results.

I hope that the text finds an interested audience, has widespread distribution, and will have some impact on the implementation of improved filtering algorithms.

Chair of Applied Statistics Prof. Dr. Hermann Singer
and Empirical Social Research
FernUniversität in Hagen
Hagen, Germany

Acknowledgments

My special thanks go to Prof. Dr. Hermann Singer from the Chair of Applied Statistics and Empirical Social Research at the FernUniversität in Hagen, who has always been at my side with advice and action. He has always encouraged and motivated me, especially in the stages where progress was scarcely apparent and promising approaches proved to be unsuccessful. This work would not have been possible without his commitment. Furthermore, I thank Prof. Dr. Wilhelm Rödder, my second examiner, for his helpful remarks during his co-chairmanship.

My colleagues, Andrea Buczek, Zulfiya Davidova, Daniela Doliwa, Marina Lorenz, Armin Müller, Bayram Oruc, and Frederik Parton, make the chair of what it is: a workplace with a professional and friendly atmosphere. You are a great team!

Last but not least I would like to thank my parents who have always believed in me and supported me in every possible way.

Hagen, Germany Dominik Ballreich
May 2017

Contents

1 **Introduction** 1
   1.1 Problem Statement and Objective 2
   1.2 Outline 3

2 **Filtering in Dynamical Systems** 5
   2.1 The General Discrete State-Space Model 6
   2.2 The Bayes Filter 6
   2.3 The Kalman Filter 10
      2.3.1 The Kalman Filter Algorithm in the Case of the Gaussian Linear Discrete State-Space Model 11
      2.3.2 The Nonlinear Kalman Filter and the Gaussian Assumption 13
   2.4 Parameter Estimation 21
      2.4.1 Maximum Likelihood Estimation 21
      2.4.2 Bayesian Parameter Estimation 22
   2.5 Conditional Filtering 31
   2.6 Stabilization of Nonlinear Kalman Filter Algorithms 42
   2.7 Treatment of Missing Data 43

3 **Deterministic Numerical Integration** 47
   3.1 One-Dimensional Deterministic Numerical Integration 48
      3.1.1 Lagrange Interpolation 48
      3.1.2 Moment Equations for the One-Dimensional Case 49
      3.1.3 Gauss Quadrature 53
      3.1.4 Clenshaw–Curtis Quadrature 62
   3.2 Multidimensional Deterministic Numerical Integration 65
      3.2.1 Stability Factor 66
      3.2.2 A Lower Bound for the Number of Abscissae 67
      3.2.3 Polynomials in d Dimensions 68
      3.2.4 Product Cubature Rules 69
      3.2.5 Moment Equations for the d-Dimensional Case 71
      3.2.6 Smolyak Cubature 80

		3.2.7	Compound Rules	88
		3.2.8	Change of Variables	89
4	**Optimization and Stabilization of Cubature Rules**			93
	4.1	Cubature Rules Based on a Least Squares Approach		93
	4.2	Construction of Stabilized Smolyak Cubature Rules		98
		4.2.1	Stabilized(1) Rules	99
		4.2.2	Stabilized(2) Rules	102
		4.2.3	Smolyak Cubature Rules with an Approximate Degree of Exactness	106
5	**Simulation Studies**			109
	5.1	The Univariate Non-Stationary Growth Model		112
	5.2	The Six-Dimensional Coordinated Turn Model		117
	5.3	The Lorenz Model		126
	5.4	The Ginzburg–Landau Model		130
6	**Results**			135
A	**The Conditional Mean**			139
B	**The Moments of the Conditional Normal Distribution**			141
C	**The Golub–Welsch Algorithm**			143
D	**Simplified Multidimensional Moment Equations**			149
Bibliography				153
Index				159

List of Figures

Fig. 2.1	Bayesian parameter estimation of a drift parameter by the example of an AR(2)-Model	23
Fig. 2.2	Bayesian parameter estimation of a diffusion parameter without initial covariance by the example of an AR(2)-Model	27
Fig. 2.3	Bayesian parameter estimation of a diffusion parameter with initial covariance by the example of an AR(2)-Model	31
Fig. 2.4	Bayesian parameter estimation of a diffusion parameter by the example of an AR(2)-Model (UCKF)	42
Fig. 2.5	Missing data and prediction	44
Fig. 3.1	Tensor products	70
Fig. 3.2	Comparison of various underlying quadrature rules and tensor product for $d = 2$	85
Fig. 3.3	Comparison of various underlying quadrature rules and tensor product for $d = 10$	86
Fig. 3.4	Comparison of various underlying quadrature rules and tensor product for $m = 13$	87
Fig. 3.5	Decomposition-levels	89
Fig. 4.1	Comparison of stability factors, $m = 13$	98
Fig. 4.2	Comparison of various underlying quadrature rules and tensor product to the stabilized rule for $m = 13$	102
Fig. 5.1	Cobweb plot	112
Fig. 5.2	The univariate non-stationary growth model	113
Fig. 5.3	Exemplary prior and posterior density of the non-stationary growth model	113
Fig. 5.4	The coordinated turn model	123
Fig. 5.5	The stochastic Lorenz model	127
Fig. 5.6	The stochastic Ginzburg–Landau model ($\alpha = -1, \beta = 1$)	131
Fig. 5.7	Time evolution of the prior density between measurements ($\alpha = -1, \beta = 1$)	132

List of Tables

Table 4.1	Cubature rules with approximate degree of exactness for the unweighted case	97
Table 4.2	Cubature rules with approximate degree of exactness in the case of the Gaussian weight function	97
Table 4.3	Comparison of stability factors in the case of the weight function $w(x) = 1$ and $m = 13$	100
Table 4.4	Comparison of stability factors in the case of the Gaussian weight function and $m = 13$	101
Table 4.5	Comparison of stabilization methods, $m = 9$	104
Table 4.6	Comparison of stabilization methods, $m = 11$	105
Table 4.7	Comparison of Smolyak-AE and Smolyak cubature rules, $m = 9$ and $m = 11$	107
Table 5.1	Comparison of MADs (univariate non-stationary growth model)	116
Table 5.2	Comparison of computation times (univariate non-stationary growth model)	116
Table 5.3	Comparison of MADs (coordinated turn model)	124
Table 5.4	Comparison of computation times (coordinated turn model)	125
Table 5.5	Comparison of MADs (Lorenz model)	129
Table 5.6	Comparison of computation times (Lorenz model)	130
Table 5.7	Comparisons of MADs and computation times (Ginzburg–Landau model)	133

List of Symbols and Abbreviations

α_l	Cubature weights
chol(Σ), lower	Lower triangular matrix of the Cholesky decomposition of Σ
$\mathbb{E}[\mathbf{y}_{t\|t-1}]$	Conditional mean of the state, given the measurements up to time $t-i$, $(i=0,1)$
$\mathbb{P}_n$	The $n+1$-dimensional polynomial vector space, spanned by the monomial basis $\{1,\ x,\ x^2,\ldots,\ x^n\}$
$Q^{m,n}$	Quadrature rule which uses n abscissas and has degree of exactness m
S^c	A set which includes c quadrature/cubature abscissae
ϑ_l	Quadrature weights
$\mathbf{y}_t$	State vector
$\mathbf{z}_t$	Measurement vector
$\mathbf{Z}^{t-i}$	Measurements up to time $t-i$, $(i=0,1)$
AE	Cubature rules with an approximate degree of exactness (based on the multidimensional moment equations)
CKF	Cubature Kalman filter of Arasaratnam and Haykin
DBF	Discrete Bayes filter
DGKP	Smolyak cubature based on delayed Gauss-Kronrod-Patterson rules
GH	Smolyak cubature based on Gauss-Hermite rules
GK	Smolyak cubature based on Genz-Keister rules
GL	Smolyak cubature based on Gauss-Legendre rules
MAD	Mean absolute deviation
NCC	Smolyak cubature based on delayed Clenshaw-Curtis rules
Smolyak-AE	Smolyak cubature rules with an approximate degree of exactness
UCKF	Unscented conditional Kalman filter of Singer
UKF	Unscented Kalman filter of Julier and Uhlmann

Chapter 1
Introduction

Research questions from various fields of science ranging from biology over engineering to economics are related to the problem of estimating latent time-dependent variables, based on noisy measurements. The associated estimation problems are called filter problems. Focusing on applications in economics, countless tasks exist which can be tackled by applying filter techniques. Examples are

- the efficient handling of missing data while forecasting time series using, for instance, autoregressive integrated moving average (ARIMA) models (cf. Box et al. 2008, pp. 544–550),
- the estimation of stochastic volatility time series (cf. Singer 2015) and
- the likelihood estimation in nonlinear and highly parametrized dynamic stochastic general equilibrium (DSGE) models (cf. Fernández-Villaverde 2010),

to mention only a few.

The foundation for the solution to filter problems is given by Bayes' theorem (Bayes and Price 1763). Various filter algorithms are based on this theorem, which therefore all belong to the important group of Bayesian filter algorithms. The theorem itself can be implemented algorithmically in many ways and is then called Bayes filter. However, the known implementations have the weak point that, due to the high complexity of the necessary calculations, they can provide satisfactory results only if the number of variables to be filtered is very small.

Given a Gaussian and linear filter problem, an exact, recursive, efficient and easy to implement algorithmic version of Bayes' theorem exists, known as the Kalman filter (Kalman 1960). Beyond this special case, the filter can also handle non-Gaussian, nonlinear problems by making certain simplifying assumptions. Due to these simplifications, the resulting filter then is only an approximation to the Bayes filter. It turns out, however, that the arising filter solutions are notably accurate with respect to the majority of filter problems. Due to its many positive attributes the Kalman filter is probably the most commonly used method for treating filter problems in research and practice. This is also reflected by the fact that in recent decades, the research activity regarding the Kalman filter has steadily increased and numerous improvements and extensions have been published.

1.1 Problem Statement and Objective

In the aforementioned filter algorithms, the variables to be estimated are represented by the so-called state vector of dimension d. The core problem here is that the execution of the algorithms repeatedly requires the calculation of d-dimensional integrals (conditional expectations). Only in rare cases analytical solutions to these integrals exist and accordingly, methods of numerical integration must be applied. The computational complexity of these numerical methods increases with the dimension of the state vector. While low-dimensional integrals can often be evaluated almost exactly within a reasonable time, from a certain dimension, depending on the chosen filter algorithm, even satisfactory approximations are hardly possible. Beyond computational complexity, stability is a crucial factor that is almost more important. An unstable method of numerical integration can lead to poor results or even to the divergence of the filter, and should therefore be avoided.

The numerical technique used in this work is known as deterministic numerical integration and also often called quadrature or, in the case of multidimensional integrals, cubature. In this method, the approximation of the considered multidimensional integrals is performed by summing up the products of specific function values $f(\chi_l)$ and weights α_l, in which the evaluation points χ_l are called abscissae.[1] An important theoretical measure of the quality of a certain cubature rule is its degree of polynomial exactness, in the following also referred to as exactness. An exactness of m implicates that arbitrary polynomials of degree up to m can be integrated exactly by applying the respective cubature rule. Based on this measure, a cubature rule is more efficient than another, if it possesses the same exactness but uses less abscissae. The other important issue, which is the stability of a cubature rule, is strongly connected to the signs of the used weights. Ideally, all weights should be positive, since otherwise amplified rounding errors may distort the results.

The efficient and stable approximation of the relevant integrals by using cubature rules is the major focus of this work. Two approaches for creating these rules are discussed, which are the so-called moment equations approach as well as the Smolyak algorithm (cf. Smolyak 1963). The Smolyak algorithm takes quadrature rules as inputs and combines them into cubature rules. The arising Smolyak cubature rules are, depending on the structure of the input quadrature rules, extremely efficient. However, a disadvantage of the Smolyak cubature rules known from literature is their poor stability. By the use of both approaches in combination with numerical optimization routines, four new kinds of cubature rules are generated which exhibit

---

[1] $\int_{I_1} \ldots \int_{I_d} f(x_1, \ldots, x_d) \, dx_d \ldots dx_1 \approx \sum_{l=1}^{n} f(\chi_l) \alpha_l$, with $I_1, \ldots, I_d \subseteq \mathbb{R}$.

significant advantages in terms of efficiency and stability compared to rules already known from literature. The main attention in the construction of the new rules is on stability and therefore on minimization of the influence of negative weights.

In order to investigate the performance of the self-created rules, four simulation studies are conducted in which the rules are combined with different filter algorithms. Of special interest is the impact of negative weights on the filter process. Therefore, the Smolyak cubature rules proposed by Heiss and Winschel, which have already been applied in the context of likelihood approximation (Heiss and Winschel 2008) and filtering (Winschel and Krätzig 2010) serve as comparison methods. These rules are most efficient for certain degrees of polynomial exactness and dimensions, but also lack stability. Thus, a comparison may reveal some insights into the impact of unstable cubature rules upon the filter performance.

1.2 Outline

The second chapter begins with the description of the elementary concepts of Bayesian filter theory. Based on these concepts, popular filter algorithms for discrete time linear and nonlinear filtering problems are derived, namely the discrete Bayes filter, the Kalman filter and the unscented Kalman filter (Julier et al. 1995). Furthermore, a large part of the chapter is devoted to parameter estimation, particularly to the Bayesian estimation of parameters. Here the fact that it is not possible to perform Bayesian estimation of diffusion parameters by using the nonlinear Kalman filter is analysed in detail. As a solution to this problem, the conditional filter of Singer (2015) is reviewed.

In the first part of the third chapter, classical approaches to one-dimensional deterministic integration are presented. The second part is devoted to multidimensional deterministic integration and focuses on the two concepts used for the computation of own cubature rules. These are the multidimensional moment equations and the Smolyak algorithm. From the moment equations, two cubature methods of exactness $m = 3$ frequently used for the purpose of filtering are derived. Those are the unscented transform of Julier et al. (1995) and the cubature rule of Arasaratnam and Haykin (2009). Additionally, the relationship between the cubature rule of Arasaratnam and the rule of Stroud and Secrest (1963) is pointed out. At the end of the chapter, a cubature method is reviewed, which is based on the decomposition of the integration domain. The associated rules are called compound rules.

The fourth chapter comprises the main research results with respect to the creation of own cubature rules. The so-called AE rules (approximate exactness), described in Sect. 4.1, are constructed based on a least squares approach to the moment equations. Due to the resulting small approximation errors, these rules have an *approximative* degree of polynomial exactness. This means that the arising cubature rules from the outset are constructed in a way that slight inaccuracies with respect to the integration of polynomials are tolerated. The weights used by the computed rules are strictly positive and therefore the level of stability is optimal. Ignoring the small approximation errors, no other rules known from literature which use strictly positive weights are more efficient than the AE rules. Furthermore, in most of cases under consideration, the new rules use less abscissae than the most efficient rules known from literature in general.

In Sect. 4.2. an algorithm for the computation of new kinds of Smolyak cubature rules is derived. This approach is used in order to generate rules which are of higher exactness and are usable for higher dimensions than the AE rules. The first two types of new Smolyak cubature rules are the stabilized(1) and stabilized(2) rules. Both kinds of rules are constructed with the focus lying mainly on stability. The stabilized(1) rules are less efficient than other rules known from literature, but significantly more stable. An extension to this approach leads to the stabilized(2) rules, which are characterized by a higher efficiency and lower stability compared to the stabilized(1) rules. Although the considerable increase of efficiency leads to a reduction of stability, the degree of stability achieved is still very high. The third kind of self-constructed Smolyak cubature rules are the Smolyak-AE rules (Smolyak cubature rules of approximate exactness). The algorithm applied is based on the idea of shrinking some of the weights towards zero. These weights then can be omitted because their influence on the cubature result is negligible. However, omitting some of the weights, even if they are very small, always leads to small inaccuracies with regard to the cubature rule. Therefore, the Smolyak-AE rules have an *approximate* degree of exactness. Neglecting the small inaccuracies, these rules are more efficient than all other rules known from literature. As far as stability is concerned, the Smolyak-AE rules outperform the Smolyak cubature rules of Heiss and Winschel.

In the fifth chapter, the performance of the self-created rules is compared to the performance of the Smolyak cubature rules proposed by Heiss and Winschel. For this purpose, the filter algorithms described in the first chapter are equipped with AE/Smolyak-AE as well as with the comparing Smolyak cubature rules, and applied within four simulation studies.

Chapter 2
Filtering in Dynamical Systems

A dynamical system describes the dynamics of a time-dependent physical or even pure mathematical phenomenon. The state of the system is represented by the state vector y_t. This vector contains the values of all variables necessary to describe the system at a certain point t in time (cf. Kalman 1963, p. 154). In the deterministic case, the time evolution of the state vector is fully predetermined by the state transition equation and possibly also exogenous variables. In the stochastic case, the dynamics of the state vector are additionally influenced by random perturbations.

The key problem in controlling the dynamical system or in reacting to its behaviour is the fact that the state itself is latent. Only noisy measurements z_t are available and, furthermore, in many cases not all but only specific components of the vector y_t are measurable. An illustrative example for this scenario is the position tracking of a vehicle. In the simplest case, the state vector y_t includes two components, namely the position and velocity of the object at time t. Only the position component is available, in the form of GPS measurements. However, these measurements are not exact and include errors. The goal of filtering is therefore to estimate the expected value of the whole state at every point in time, based on noisy measurements.

Treating the quantities *state* and *measurement* as a joint system, this system is formulated as a *state-space model*. With respect to the time framework, a state-space model can be presented in various ways. If the state is modeled in continuous time while the measurements are of discrete nature, the so-called *continuous/discrete* state-space model emerges (cf. Jazwinski 2007, p. 163). Accordingly, also the continuous/continuous and the discrete/discrete state-space model exists. In this work, the discrete/discrete approach will be used.

To simplify the notation, the measurement series $\{z_t, z_{t-1}, \ldots, z_1\}$ which includes all measurements available until time t will be denoted by Z^t. Therefore, Z^t can also be split in several parts, for example, $Z^t = \{z_t, Z^{t-1}\}$.

2.1 The General Discrete State-Space Model

The general discrete state-space model (cf. Jazwinski 2007, p. 174) can be formulated as

$$y_t = f(y_{t-1}, x_t, \phi) + g(y_{t-1}, x_t, \phi)\epsilon_t \qquad (2.1)$$

$$z_t = h(y_t, x_t, \phi) + \delta_t. \qquad (2.2)$$

The state-space model comprises two components: the recursive state equation (2.1) and the measurement equation (2.2). Consisting of a drift function and a diffusion function, the recursive state equation represents the dynamics of the unobservable $d \times 1$ state vector y_t, assuming that the underlying stochastic process is autoregressive of order one and possesses the Markov property.[1]

In contrast, the measurement equation is a function of the *current* state, superimposed by an additive white noise sequence δ_t. The random noise vectors ϵ_t and δ_t of dimension $r \times 1$ and $m \times 1$ are normally distributed with $\epsilon_t \sim \mathcal{N}(0, I)$ and $\delta_t \sim \mathcal{N}(0, R(x_t, \phi))$, respectively. Furthermore, both vectors are defined to be independent, so that $\mathbb{E}\left[\epsilon_i \delta_j'\right] = 0 \ \forall \ i, j$. It has to be mentioned that the derivations of nonlinear filter algorithms, which will be stated in the further course of this work, are independent of the distribution of the error terms. Nevertheless, the use of Gaussian errors is the standard case. Regarding the diffusion function $g(y_{t-1}, x_t, \phi)$ it holds that $g(y_{t-1}, x_t, \phi) g(y_{t-1}, x_t, \phi)' = \Omega(y_{t-1}, x_t, \phi)$. Additionally, both equations depend on a $p \times 1$ parameter vector ϕ and, optionally, a $q \times 1$ vector of deterministic control variables x_t. In addition, the drift function $f : \mathbb{R}^d \times \mathbb{R}^p \times \mathbb{R}^q \to \mathbb{R}^d$, the diffusion function $g : \mathbb{R}^d \times \mathbb{R}^p \times \mathbb{R}^q \to \mathbb{R}^d \times \mathbb{R}^r$, the measurement function $h : \mathbb{R}^d \times \mathbb{R}^p \times \mathbb{R}^q \to \mathbb{R}^m$ and the measurement error covariance $R(x_t, \phi)$ may also depend on past measurements Z^{t-1}. The distribution of the initial condition $p(y_0)$ is assumed to be given and to be independent of ϵ_0 (cf. Singer 2008, p. 180).

2.2 The Bayes Filter

The minimum variance unbiased estimator for the latent state y_t given the measurements Z^t is the conditional mean (cf. Jazwinski 2007, pp. 145–150)[2]

$$\mathbb{E}\left[y_t | Z^t\right] = \int_{-\infty}^{\infty} y_t p(y_t | Z^t) \, dy_t. \qquad (2.3)$$

---

[1] $p(y_t | y_{t-1}, y_{t-2}, \ldots, y_1) = p(y_t | y_{t-1})$ (cf. Klebaner 2005, p. 67).
[2] A more detailed explanation can be found in Appendix A.

2.2 The Bayes Filter

For the evaluation of this and other conditional moments, the conditional density $p(y_t|Z^t)$ is needed. To obtain a recursive form for every time step, the condition Z^t is divided into two parts,

$$p(y_t|Z^t) = p(y_t|z_t, Z^{t-1}). \quad (2.4)$$

This density is derived by using Bayes' theorem (cf. Singer 2011, p. 13):

$$\begin{aligned} p(y_t|z_t, Z^{t-1}) &= \frac{p(z_t, Z^{t-1}, y_t)}{p(z_t, Z^{t-1})} \\ &= \frac{p(z_t|y_t, Z^{t-1}) p(y_t, Z^{t-1})}{p(z_t, Z^{t-1})} \\ &= \frac{p(z_t|y_t, Z^{t-1}) p(y_t|Z^{t-1}) p(Z^{t-1})}{p(z_t|Z^{t-1}) p(Z^{t-1})} \\ &= \frac{p(z_t|y_t, Z^{t-1}) p(y_t|Z^{t-1})}{p(z_t|Z^{t-1})}. \end{aligned} \quad (2.5)$$

The emerging Bayes filter consists of two parts:

1. *Time update (a priori density)*

$$p(y_t|Z^{t-1}) = \int_{-\infty}^{\infty} p(y_t|y_{t-1}) p(y_{t-1}|Z^{t-1}) dy_{t-1} \quad (2.6)$$

The a priori density (PR_t) is constructed by integrating over the product of transition density (TD_t) $p(y_t|y_{t-1})$ and a posteriori density (PO_t) $p(y_{t-1}|Z^{t-1})$. Equation (2.6) is called *Chapman–Kolmogorov Equation*. After a new measurement z_t, the measurement update follows.

2. *Measurement update (a posteriori density)*

$$p(y_t|z_t, Z^{t-1}) = \frac{p(z_t|Z^{t-1}, y_t) \int_{-\infty}^{\infty} p(y_t|y_{t-1}) p(y_{t-1}|Z^{t-1}) dy_{t-1}}{p(z_t|Z^{t-1})} \quad (2.7)$$

The a priori density is multiplied by the measurement density (MD_t) $p(z_t|Z^{t-1}, y_t)$. Normalizing the product by the likelihood (L) $p(z_t|Z^{t-1})$ yields the a posteriori density (PO_t).

The discrete version of the Bayes filter can be implemented by approximating the needed densities by Riemann sums. Each dimension of the state is discretized into n_j sub-intervals from a_j to b_j with length $\Delta_j = \frac{b_j - a_j}{n_j}$. The ith coordinate of dimension j then is given by

$$k_{i_j} = a_j + \Delta_j \cdot i_j, \; i_j = 0, \ldots, n_j, \; j = 1, \ldots, d \tag{2.8}$$

and the approximation to the value of an arbitrary integral

$$\mathcal{I}[f] = \int_{a_1}^{b_1} \ldots \int_{a_d}^{b_d} f(x_1, \ldots, x_d) \, dx_d \ldots dx_1 \tag{2.9}$$

reads

$$\sum_{i_1=0}^{n_1} \ldots \sum_{i_d=0}^{n_d} f(k_{i_1}, \ldots, k_{i_d}) \Delta_1 \ldots \Delta_d. \tag{2.10}$$

Thus, the approximation is carried out by evaluating the function on a tensor product constructed on the basis of the one-dimensional discretizations and summing over the weighted values. In shorthand notation this can be formulated as follows:

$$\mathcal{I}[f] \approx \sum_{l=1}^{n} f(\boldsymbol{\chi}_l) \alpha_l, \; \text{with } n = \prod_{j=1}^{d} (n_j + 1). \tag{2.11}$$

The limits a_j and b_j, $j = 1, \ldots, d$, are set at the beginning of the filtering procedure and restrict the actually infinite domain of integration to a region where the state y_t resides. This region can, for example, be approached on the basis of simulations of the state-space model. The use of grid-based methods poses a variety of problems in the practical application. A rough approximation of the densities can lead to poor results and even to the divergence of the algorithm. Therefore, the used grid must be very dense, which leads to a high computational intensity already for dimensions $d > 2$. A basic algorithm which uses a fixed grid will be now presented. A more refined algorithm, in which the grid changes in every iteration based on additional information, has been proposed by Tanizaki (cf. Tanizaki 2010, pp. 78–85).

Apart from the presented algorithm other numerical methods exist by which the required densities can be approximated. A widely used method is the particle filter and its extensions, which employs Monte-Carlo-Integration (cf. Arulampalam et al. 2002). Nevertheless, all existing filter algorithms have in common that they, in many cases, are insufficient, numerically unstable and computationally inefficient. This problem is amplified with increasing dimensionality of the state.

2.2 The Bayes Filter

Algorithm 1 The discrete Bayes filter

1: **procedure** DBF
2:   *Initialization*:
3:   μ; Σ; $\chi = [\chi_1, \ldots, \chi_n]$; $\alpha = [\alpha_1, \ldots, \alpha_l]$; $\mathbf{PR}_1 = [\mathcal{N}(\chi_1; \mu, \Sigma), \ldots, \mathcal{N}(\chi_n; \mu, \Sigma)]$
4: ―――――――――――――――――――――――
5:   **for** $t = 1$ **do**
6:     *Measurement update*
7:   **end for**
8:   **for** $t = 2 : T$ **do**
9:     *Time update*:
10:    $Y_l = f(\chi_l, x_t, \phi)$, $l = 1, \ldots, n$
11:    $Z_l = h(\chi_l, x_t, \phi)$, $l = 1, \ldots, n$
12:    $\Omega_l = \Omega(\chi_l, x_t, \phi)$, $l = 1, \ldots, n$
13:    **for** $i = 1 : n$ **do**
14:      **for** $j = 1 : n$ **do**
15:        $\mathbf{TD}_{t,ij} = \mathcal{N}(\chi_i; Y_j, \Omega_j)$
16:      **end for**
17:    **end for**
18:    $\mathbf{PR}_t = \mathbf{TD}_t \cdot (\mathbf{PO}_{t-1} \odot \alpha)'$
19:    *Measurement update*:
20:    $Z_l = h(\chi_l, x_t, \phi)$, $l = 1, \ldots, n$
21:    $R_t = R(x_t, \phi)$
22:    $\mathbf{MD}_t = [\mathcal{N}(z_t; Z_1, R_t), \ldots, \mathcal{N}(z_t; Z_n, R_t)]$
23:    $L(z_t) = \mathbf{MD}_t \cdot (\mathbf{PR}_t \odot \alpha)'$
24:    $\mathbf{PO}_t = \frac{\mathbf{MD}_t \odot \mathbf{PR}_t}{L(z_t)}$
25:    $\mathbb{E}[y_t | Z^t] = \chi \cdot (\mathbf{PO}_t \odot \alpha)'$
26:    $\chi^C = \chi - (\mu_{y,t|t} \cdot [1, 1, \ldots, 1])$
27:    $\mathbb{V}[y_t | Z^t] = (\chi^C \odot (\mathbf{PO}_t \odot \alpha)') \chi^{C'}$
28: **end for**
29: **end procedure**[3]

---

[3] $[a\ b] \odot [c\ d] := [ac\ bd]$ (Element-wise product)

2.3 The Kalman Filter

An important special case of the general discrete state-space model is the Gaussian linear discrete state-space model

$$y_t = A_{t-1} y_{t-1} + x_t + \epsilon_t$$
$$z_t = H_t y_t + d_t + \delta_t. \tag{2.12}$$

The deterministic matrices A_{t-1} and H_t are of dimensions $d \times d$ and $m \times d$ and the random noise vectors ϵ_t and δ_t are normally distributed with $\epsilon_t \sim \mathcal{N}(\mathbf{0}, \mathbf{\Omega}_t)$ and $\delta_t \sim \mathcal{N}(\mathbf{0}, R_t)$, respectively.

From the fact that the random vectors y_t and z_t are both normally distributed it follows from Bayes' theorem that the density $p(y_t|z_t)$ is conditionally normal. The moments, through which the distribution is defined, are given by the *theorem on normal correlation* (cf. Liptser and Shiryaev 2001, p. 61):

The moments of the conditional multivariate Normal Distribution

$$\mathbb{E}[y_t|z_t] = \mathbb{E}[y_t] + \text{Cov}[y_t, z_t] \, \mathbb{V}[z_t]^{-1} (z_t - \mathbb{E}[z_t]) \tag{2.13}$$

$$\mathbb{V}[y_t|z_t] = \mathbb{V}[y_t] - \text{Cov}[y_t, z_t] \, \mathbb{V}[z_t]^{-1} \text{Cov}[y_t, z_t]' \tag{2.14}$$

The measurement z_t is included in the calculation of the conditional mean in form of a linear regression from y_t on z_t. The term $\text{Cov}[y_t, z_t] \, \mathbb{V}[z_t]^{-1}$ is called Kalman gain (K_t) and can be interpreted as regression coefficient. Concerning the conditional variance a notable feature is that the unconditional variance $\mathbb{V}[y_t]$ is always greater than or equal to the conditional variance $\mathbb{V}[y_t|z_t]$. The inclusion of the information z_t therefore has a variance reducing effect. Rewriting equation (2.13) into

$$\mathbb{E}[y_t|z_t] = \underbrace{\mathbb{E}[y_t] - \underbrace{\text{Cov}[y_t, z_t] \, \mathbb{V}[z_t]^{-1}}_{K_t} \mathbb{E}[z_t]}_{\text{Intercept}} + \underbrace{\text{Cov}[y_t, z_t] \, \mathbb{V}[z_t]^{-1}}_{K_t} z_t \tag{2.15}$$

shows the similarities to the ordinary linear regression (scalar case, for simplicity):

$$\hat{y} = a + bx$$
$$b = \frac{\text{Cov}[y, x]}{\mathbb{V}[x]}$$
$$a = \mathbb{E}[y] - b\mathbb{E}[x] \tag{2.16}$$
$$\Rightarrow \hat{y} = \underbrace{\mathbb{E}[y] - \underbrace{\frac{\text{Cov}[y, x]}{\mathbb{V}[x]}}_{\text{Regression coefficient}} \mathbb{E}[x]}_{\text{Intercept}} + \underbrace{\frac{\text{Cov}[y, x]}{\mathbb{V}[x]}}_{\text{Regression coefficient}} x.$$

2.3 The Kalman Filter

In the case, where y_t is conditional upon two variables $Z^t = \{z_t, Z^{t-1}\}$, the moments read

$$\mathbb{E}\left[y_t|Z^t\right] = \mathbb{E}\left[y_t|Z^{t-1}\right] + \text{Cov}\left[y_t, z_t| Z^{t-1}\right]\mathbb{V}\left[z_t|Z^{t-1}\right]^{-1} \left(z_t - \mathbb{E}\left[z_t|Z^{t-1}\right]\right) \tag{2.17}$$

$$\mathbb{V}\left[y_t|Z^t\right] = \mathbb{V}\left[y_t|Z^{t-1}\right] - \text{Cov}\left[y_t, z_t| Z^{t-1}\right]\mathbb{V}\left[z_t|Z^{t-1}\right]^{-1} \text{Cov}\left[y_t, z_t|Z^{t-1}\right]'. \tag{2.18}$$

These equations are the exact version of the Bayes filter (Sect. 2.2) in the case of a Gaussian linear discrete state-space model and represent the foundation of the Kalman filter (Kalman 1960).

2.3.1 The Kalman Filter Algorithm in the Case of the Gaussian Linear Discrete State-Space Model

In the next step, the formulas for the conditional moments will be derived. To simplify the notation, the following abbreviations will be used from now on $(i = 0, 1)$:

$$\mu_{y,t|t-i} = \mathbb{E}\left[y_t|Z^{t-i}\right] \tag{2.19}$$

$$\mu_{z,t|t-i} = \mathbb{E}\left[z_t|Z^{t-i}\right] \tag{2.20}$$

$$\Sigma_{yy,t|t-i} = \mathbb{V}\left[y_t|Z^{t-i}\right] \tag{2.21}$$

$$\Sigma_{zz,t|t-i} = \mathbb{V}\left[z_t|Z^{t-i}\right] \tag{2.22}$$

$$\Sigma_{yz,t|t-i} = \text{Cov}\left[y_t, z_t| Z^{t-i}\right]. \tag{2.23}$$

Inserting the equations of the Gaussian linear discrete state-space model (2.12), the conditional moments read as follows:

$$\mu_{y,t|t-1} = \mathbb{E}\left[A_{t-1}y_{t-1} + x_{t-1} + \epsilon_t|Z^{t-1}\right]$$
$$= A_{t-1}\mu_{y,t-1|t-1} + x_{t-1} \tag{2.24}$$

$$\Sigma_{yy,t|t-1} = \mathbb{V}\left[A_{t-1}y_{t-1} + x_{t-1} + \epsilon_t|Z^{t-1}\right]$$
$$= A_{t-1}\Sigma_{y,t-1|t-1}A'_{t-1} + \Omega_t \tag{2.25}$$

$$\begin{aligned}\boldsymbol{\mu}_{z,t|t-1} &= \mathbb{E}\left[\boldsymbol{H}_t\boldsymbol{y}_t + \boldsymbol{d}_t + \boldsymbol{\epsilon}_t|\boldsymbol{Z}^{t-1}\right] \\ &= \boldsymbol{H}_t\boldsymbol{\mu}_{y,t|t-1} + \boldsymbol{d}_t \end{aligned} \quad (2.26)$$

$$\begin{aligned}\boldsymbol{\Sigma}_{zz,t|t-1} &= \mathbb{V}\left[\boldsymbol{H}_t\boldsymbol{y}_t + \boldsymbol{d}_t + \boldsymbol{\delta}_t|\boldsymbol{Z}^{t-1}\right] \\ &= \boldsymbol{H}_t\boldsymbol{\Sigma}_{yy,t-1|t-1}\boldsymbol{H}_t' + \boldsymbol{R}_t \end{aligned} \quad (2.27)$$

$$\begin{aligned}\boldsymbol{\Sigma}_{yz,t|t-1} &= \mathbb{E}\left[(\boldsymbol{y}_t - \boldsymbol{\mu}_{y,t|t-1})\left(\boldsymbol{H}_t\boldsymbol{y}_t + \boldsymbol{d}_t + \boldsymbol{\delta}_t - \boldsymbol{\mu}_{z,t|t-1}\right)|\boldsymbol{Z}^{t-1}\right] \\ &= \boldsymbol{\Sigma}_{yy,t|t-1}\boldsymbol{H}_t'. \end{aligned} \quad (2.28)$$

Using the derived expressions, the conditional moments given by the theorem on normal correlation (2.3) can be calculated. As an example, the Kalman gain, $\boldsymbol{K}_t = \boldsymbol{\Sigma}_{yz,t|t-1}\boldsymbol{\Sigma}_{zz,t|t-1}^{-1}$, reads

$$\boldsymbol{K}_t = \boldsymbol{\Sigma}_{yy,t|t-1}\boldsymbol{H}_t'\boldsymbol{\Sigma}_{zz,t|t-1}^{-1}. \quad (2.29)$$

Consequently, the conditional mean, $\boldsymbol{\mu}_{y,t|t} = \boldsymbol{\mu}_{y,t|t-1} + \boldsymbol{K}_t\left(\boldsymbol{z}_t - \boldsymbol{\mu}_{z,t|t-1}\right)$, results in the formula

$$\boldsymbol{\mu}_{y,t|t} = \boldsymbol{\mu}_{y,t|t-1} + \boldsymbol{\Sigma}_{yy,t|t-1}\boldsymbol{H}_t'\boldsymbol{\Sigma}_{zz,t|t-1}^{-1}\left(\boldsymbol{z}_t - \boldsymbol{\mu}_{z,t|t-1}\right). \quad (2.30)$$

Furthermore, using the Kalman gain $\boldsymbol{K}_t = \boldsymbol{\Sigma}_{yz,t|t-1}\boldsymbol{\Sigma}_{zz,t|t-1}^{-1}$ and the expression

$$\boldsymbol{\Sigma}_{yz,t|t-1} = \boldsymbol{\Sigma}_{yy,t|t-1}\boldsymbol{H}_t', \quad (2.31)$$

it follows that the conditional variance, $\boldsymbol{\Sigma}_{yy,t|t} = \boldsymbol{\Sigma}_{yy,t|t-1} - \boldsymbol{\Sigma}_{yz,t|t-1}\boldsymbol{\Sigma}_{zz,t|t-1}^{-1}\boldsymbol{\Sigma}_{yz,t|t-1}'$, can be expressed as

$$\begin{aligned}\boldsymbol{\Sigma}_{yy,t|t} &= \boldsymbol{\Sigma}_{yy,t|t-1} - \boldsymbol{K}_t\boldsymbol{H}_t\boldsymbol{\Sigma}_{yy,t|t-1} \\ &= (\boldsymbol{I} - \boldsymbol{K}_t\boldsymbol{H}_t)\boldsymbol{\Sigma}_{yy,t|t-1}. \end{aligned} \quad (2.32)$$

With these information, the Kalman filter algorithm can be stated.

2.3 The Kalman Filter

Algorithm 2 The Kalman filter algorithm for Gaussian linear discrete state-space models

1: **procedure** KF
2:   *Initialization*:
3:   $\mu_{y,1|0}$; $\Sigma_{yy,1|0}$
4:   ⸻
5:   **for** $t = 1$ **do**
6:      *Measurement update*
7:   **end for**
8:   **for** $t = 2 : T$ **do**
9:      *Time update*:
10:     $\mu_{y,t|t-1} = A_{t-1}\mu_{y,t-1|t-1} + x_{t-1}$
11:     $\Sigma_{yy,t|t-1} = A_{t-1}\Sigma_{yy,t-1|t-1}A'_{t-1} + \Omega_{t-1}$
12:     *Measurement update*:
13:     $\mu_{z,t|t-1} = H_t\mu_{y,t|t-1} + d_t$
14:     $\Sigma_{zz,t|t-1} = H_t\Sigma_{yy,t|t-1}H'_t + R_t$
15:     $\nu_t = z_t - \mu_{z,t|t-1}$
16:     $K_t = \Sigma_{yy,t|t-1}H'_t\Sigma_{zz,t|t-1}^{-1}$
17:     $\mu_{y,t|t} = \mu_{y,t|t-1} + K_t\nu_t$
18:     $\Sigma_{yy,t|t} = (I - K_tH_t)\Sigma_{yy,t|t-1}$
19:   **end for**
20: **end procedure**

2.3.2 The Nonlinear Kalman Filter and the Gaussian Assumption

In contrast to the Gaussian linear case, the conditional distributions of the Bayes filter cannot be determined analytically if the state-space model is nonlinear. For practical applications, an approximation by the Gaussian assumption

$$\begin{aligned} p\left(y_t|Z^{t-1}\right) &\approx \mathcal{N}\left(\mathbb{E}\left[y_t|Z^{t-1}\right], \mathbb{V}\left[y_t|Z^{t-1}\right]\right) \\ p\left(y_t|Z^t\right) &\approx \mathcal{N}\left(\mathbb{E}\left[y_t|Z^t\right], \mathbb{V}\left[y_t|Z^t\right]\right) \end{aligned} \quad (2.33)$$

is often sufficient. This simplification, again, leads to the theorem on normal correlation (2.3) and represents the starting point for the formulation of the nonlinear

Kalman filter. Due to the use of the Gaussian assumption, these filters are also often referred to as *Gaussian filter*. The necessary conditional moments in the case of the general Gaussian discrete state-space model will now be given in integral form. To shorten the notation, the function arguments are neglected and the multiple integrals $\int_{-\infty}^{\infty} \int_{-\infty}^{\infty} \cdots \int_{-\infty}^{\infty} f(x) \, d(x)$ are abbreviated by $\int_{-\infty}^{\infty} f(x) \, d(x)$ in the further course.

$$\begin{aligned}
\mu_{y,t|t-1} &= \mathbb{E}\left[f + g\epsilon_t | Z^{t-1}\right] \\
&= \mathbb{E}\left[f | Z^{t-1}\right] \\
&= \int_{-\infty}^{\infty} f \cdot p\left(y_{t-1} | Z^{t-1}\right) dy_{t-1}
\end{aligned} \quad (2.34)$$

$$\begin{aligned}
\Sigma_{yy,t|t-1} &= \mathbb{V}\left[f + g\epsilon_t | Z^{t-1}\right] = \mathbb{V}\left[f | Z^{t-1}\right] + \mathbb{V}\left[g\epsilon_t | Z^{t-1}\right] \\
&= \mathbb{V}\left[f | Z^{t-1}\right] + \mathbb{E}\left[gg' | Z^{t-1}\right] \mathbb{V}\left[\epsilon_t\right] \\
&= \mathbb{V}\left[f | Z^{t-1}\right] + \mathbb{E}\left[\Omega | Z^{t-1}\right] \\
&= \int_{-\infty}^{\infty} \left(f - \mu_{y,t|t-1}\right)\left(f - \mu_{y,t|t-1}\right)' p\left(y_{t-1} | Z^{t-1}\right) dy_{t-1} \\
&\quad + \int_{-\infty}^{\infty} \Omega p\left(y_{t-1} | Z^{t-1}\right) dy_{t-1}
\end{aligned} \quad (2.35)$$

$$\mu_{z,t|t-1} = \int_{-\infty}^{\infty} h \cdot p\left(y_t | Z^{t-1}\right) dy_t \quad (2.36)$$

$$\Sigma_{zz,t|t-1} = \int_{-\infty}^{\infty} \left(h - \mu_{z,t|t-1}\right)\left(h - \mu_{z,t|t-1}\right)' \cdot p\left(y_t | Z^{t-1}\right) dy_t + R_t \quad (2.37)$$

$$\Sigma_{yz,t|t-1} = \int_{-\infty}^{\infty} \left(y_t - \mu_{y,t|t-1}\right)\left(h - \mu_{z,t|t-1}\right)' \cdot p\left(y_t | Z^{t-1}\right) dy_t \quad (2.38)$$

The analytical evaluation of the above integrals is, in the majority of cases, not possible. This is the reason why in the last decades, several numerical methods for approximation purposes have been implemented. Because the variety of different nonlinear Kalman filter algorithms is constantly increasing, only a selection of the most important ones is mentioned at this point:

- Linearization—Extended Kalman filter (Schmidt 1966)
- Monte-Carlo-Integration—Ensemble Kalman filter (Evensen 1994, 2003)

2.3 The Kalman Filter

- Numerical Integration
  - Unscented Kalman filter (Julier et al. 1995)
  - Gauss–Hermite Kalman filter (Ito and Xiong 2000)
  - Cubature Kalman filter (Arasaratnam and Haykin 2009)
  - Smolyak Kalman filter (Winschel and Krätzig 2010)

Justification of the Theorem on Normal Correlation in the Nonlinear Case
With regard to the filtering of nonlinear state-space models the question arises, if the use of the theorem on normal correlation (2.3) is still appropriate. This question will now be discussed using the example of conditional mean. The following derivation will show that under some weak assumptions, the conditional mean given by the theorem on normal correlation is the best linear estimator for y_t, given the measurements Z^t, independent of the distributions of the noise vectors ϵ_t and δ_t.[4]

At first, a linear discrete model has to be defined which uses the available information to estimate y_t in every time step:

$$\mu_{t|t} = \overline{K}_t \mu_{y,t|t-1} + K_t \left(z_t - \mu_{z,t|t-1} \right). \tag{2.39}$$

Therefore, the filter value at time t is constructed as the weighted combination of the prognosis at time t and the measurement forecast error at time t. Defining the filter error vector $\psi_t = \left(y_t - \overline{K}_t \mu_{y,t|t-1} - K_t \left(z_t - \mu_{z,t|t-1} \right) \right)$, the goal is now to minimize the trace of the filter error covariance, or in other words, to minimize the length of ψ_t. This leads to the to quadratic loss-function

$$f\left(\overline{K}_t, K_t\right) = \mathrm{tr}\left[\mathbb{E}\left[\psi_t \psi_t'\right]\right] \tag{2.40}$$

which has to be minimized with respect to the matrices $\overline{K}_t$ and K_t. Calculating the derivative $\frac{\partial f}{\partial \overline{K}_t}$ and setting it equal to zero yields

$$-2\mathbb{E}\left[\left(y_t - \overline{K}_t \mu_{y,t|t-1} - K_t \left(z_t - \mu_{z,t|t-1}\right)\right) \mu_{y,t|t-1}'\right] = \mathbf{0}. \tag{2.41}$$

Adding and subtracting y_t leads to

$$-2\mathbb{E}\left[\left(y_t - \overline{K}_t \mu_{y,t|t-1} + y_t - y_t - K_t \left(z_t - \mu_{z,t|t-1}\right)\right) \mu_{y,t|t-1}'\right] = \mathbf{0} \tag{2.42}$$

and it follows that

$$\mathbb{E}\left[\left(y_t - K_t \left(z_t - \mu_{z,t|t-1}\right)\right) \mu_{y,t|t-1}'\right] = \overline{K}_t \mathbb{E}\left[\left(\mu_{y,t|t-1} + y_t - y_t\right) \mu_{y,t|t-1}'\right]. \tag{2.43}$$

---

[4] For a derivation of the linear discrete Kalman filter, based on a least squares approach, see Gelb (Gelb 1974, pp. 107–113).

Expansion of the right-hand side, results to

$$\overline{K}_t \mathbb{E}\left[\left(\mu_{y,t|t-1} - y_t\right) \mu'_{y,t|t-1}\right] + \overline{K}_t \mathbb{E}\left[y_t \mu'_{y,t|t-1}\right]. \quad (2.44)$$

The use of the law of total expectation yields

$$\begin{aligned}
&\mathbb{E}\left[\left(\mu_{y,t|t-1} - y_{t-1}\right) \mu'_{y,t|t-1}\right] \\
&= \mathbb{E}\left[\mathbb{E}\left[\left(\mu_{y,t|t-1} - y_t\right) \mu'_{y,t|t-1} | Z^{t-1}\right]\right] \\
&= \mathbb{E}\left[\mu_{y,t|t-1} \mathbb{E}\left[\left(\mu_{y,t|t-1} - y_t\right) | Z^{t-1}\right]'\right] \\
&= 0,
\end{aligned} \quad (2.45)$$

and one arrives at the following expression:

$$\mathbb{E}\left[\left(y_t - K_t\left(z_t - \mu_{z,t|t-1}\right)\right) \mu'_{y,t|t-1}\right] = \overline{K}_t \mathbb{E}\left[y_t \mu'_{y,t|t-1}\right], \quad (2.46)$$

from which follows that

$$\begin{aligned}
\overline{K}_t &= \frac{\mathbb{E}\left[\left(y_t - K_t\left(z_t - \mu_{z,t|t-1}\right)\right) \mu'_{y,t|t-1}\right]}{\mathbb{E}\left[y_t \mu'_{y,t|t-1}\right]} \\
&= \frac{\mathbb{E}\left[y_t \mu'_{y,t|t-1}\right] - K_t \mathbb{E}\left[\left(z_t - \mu_{z,t|t-1}\right) \mu'_{y,t|t-1}\right]}{\mathbb{E}\left[y_t \mu'_{y,t|t-1}\right]}.
\end{aligned} \quad (2.47)$$

Again, applying the law of total expectation, it becomes clear that

$$\begin{aligned}
&K_t \mathbb{E}\left[\left(z_t - \mu_{z,t|t-1}\right) \mu'_{y,t|t-1}\right] \\
&= K_t \mathbb{E}\left[\mathbb{E}\left[\left(z_t - \mu_{z,t|t-1}\right) \mu'_{y,t|t-1} | Z^{t-1}\right]\right] \\
&= K_t \mathbb{E}\left[\mu_{y,t|t-1} \mathbb{E}\left[\left(z_t - \mu_{z,t|t-1}\right) | Z^{t-1}\right]'\right] \\
&= 0,
\end{aligned} \quad (2.48)$$

and that therefore $\overline{K}_t = I$. Inserting $\overline{K}_t$ into (2.40), the loss-function changes to

$$f(K_t) = \mathbb{E}\left[\left(y_t - \mu_{y,t|t-1} - K_t\left(z_t - \mu_{z,t|t-1}\right)\right) \right. \\
\left. \cdot \left(y_t - \mu_{y,t|t-1} - K_t\left(z_t - \mu_{z,t|t-1}\right)\right)'\right]. \quad (2.49)$$

2.3 The Kalman Filter

Calculating the derivative $\frac{\partial f}{\partial K_t}$ and setting it equal to zero yields

$$2\mathbb{E}\left[\left(y_t - \mu_{y,t|t-1} - K_t\left(z_t - \mu_{z,t|t-1}\right)\right)\left(z_t - \mu_{z,t|t-1}\right)'\right] = 0 \qquad (2.50)$$

and it follows the well-known expression for K_t, the Kalman gain

$$K_t = \mathbb{E}\left[\left(y_t - \mu_{y,t|t-1}\right)\left(z_t - \mu_{z,t|t-1}\right)'\right]\mathbb{E}\left[\left(z_t - \mu_{z,t|t-1}\right)\left(z_t - \mu_{z,t|t-1}\right)'\right]^{-1}$$
$$= \mathrm{Cov}\left[y_t, z_t | Z^{t-1}\right]\mathbb{V}\left[z_t | Z^{t-1}\right]^{-1}. \qquad (2.51)$$

Thus, the estimator (2.39) takes on the form of the conditional mean given by the theorem on normal correlation which is therefore the optimal linear estimator for the latent state y_t, given the measurements Z^t. Unfortunately, the exact calculation of the estimator is not possible because the necessary conditional densities $p\left(y_t | Z^t\right)$ and $p\left(y_t | Z^{t-1}\right)$ are unknown. As already mentioned, this problem is circumvented by applying the Gaussian assumption $p\left(y_t | Z^{t-1}\right) \approx \mathcal{N}\left(\mu_{y,t|t-1}, \Sigma_{yy,t|t-1}\right)$ and $p\left(y_t | Z^t\right) \approx \mathcal{N}\left(\mu_{y,t|t}, \Sigma_{yy,t|t}\right)$.

On closer inspection and under the assumptions made regarding the initial linear estimation model, the nonlinear Kalman filter therefore yields approximations in three ways. These can be summarized as follows:

1. The use of the theorem on normal correlation "only" leads to the best *linear* estimator.
2. This estimator cannot be calculated exactly, because the necessary densities are unknown. To circumvent this problem, the Gaussian assumption is applied.
3. The resulting Gaussian integrals in almost no case can be evaluated analytically and thus have to be approximated by numerical methods.

Despite of the simplifications done and the necessity to use numerical approximation methods, nonlinear Kalman filter algorithms have proven to deliver stable and reliable estimation results in practical applications. A widely used algorithm is the unscented Kalman filter, which will be described in the following section.

The Unscented Kalman Filter The unscented Kalman filter (UKF) of Julier et al. (1995) is based on deterministic numerical integration. In order not to anticipate contents of the following chapter, the filter will be derived in the sense of Singer (2006a), using a statistical approach. By means of employing $2d + 1$ abscissae χ_l

and weights a_l, the singular density of a (multivariate) random variable x can be written as follows:

$$p(x) \approx \sum_{l=-d}^{d} a_l \delta(x - \chi_l).{}^5 \qquad (2.52)$$

As an example, the approximated expected value of a function $f(x)$ then can be derived straightforward as

$$\mathbb{E}[f(x)] = \int_{-\infty}^{\infty} f(x) \cdot p(x)\,dx \approx \int_{-\infty}^{\infty} f(x) \cdot \sum_{l=-d}^{d} a_l \delta(x - \chi_l)\,dx = \sum_{l=-d}^{d} a_l f(\chi_l). \qquad (2.53)$$

The method for the approximation of nonlinear functions of the random variable x now to be introduced is referred to as *unscented transform* by Julier and Uhlman. In order to provide a certain degree of accuracy, the proposed abscissae and weights are computed in a way so that they are capable of reproducing the mean and the variance of the random variable x. Therefore, the unknown density $p(x)$ is approximated by its first two moments.

The proposed abscissae are of the form

$$\chi_0 = \mu$$
$$\chi_l = \mu + \sqrt{d + \kappa}\,\Gamma_{.l},\ l = 1, \ldots, d \qquad (2.54)$$
$$\chi_{-l} = \mu - \sqrt{d + \kappa}\,\Gamma_{.l},\ l = 1, \ldots, d,$$

in which Γ is the lower triangular matrix of the Cholesky decomposition (cf. Bronštejn et al. 2005, p. 920) of Σ. So $\Gamma \Gamma' = \Sigma$. The expression $\Gamma_{.l}$ represents the lth column of the lower triangular matrix.

The weights are calculated as follows:

$$a_0 = \frac{\kappa}{(d + \kappa)}$$
$$a_l = \frac{1}{(2(d + \kappa))} = a_{-l},\ l = 1, \ldots, d \qquad (2.55)$$
$$\sum_{l=-d}^{d} a_l = 1.$$

---

[5]$\delta(x)$ is the Dirac delta "function" (cf. Tang 2007, p. 725), with

$$\delta(x) = \begin{cases} +\infty, & \text{if } x = 0 \\ 0, & \text{if } x \neq 0 \end{cases}, \text{ and } \int_{-\infty}^{\infty} \delta(x)\,dx = 1.$$

2.3 The Kalman Filter

The parameter κ determines the spread of the abscissae from the origin and can be chosen at will. The optimal value for κ is dependent on the particular integration problem and generally unknown in practical applications. Thus, the use of the parameter is questionable. The requirement to be met is that $\mathbb{E}[x] = \sum_{l=-d}^{d} \chi_l a_l$ and $\mathbb{V}[x] = \sum_{l=-d}^{d} (\chi_l - \mu)(\chi_l - \mu)' a_l$. Substituting $\sqrt{d+\kappa}$ for q, $q \geq \sqrt{d}$, it turns out that

$$\mathbb{E}[x] = \frac{d\mu + (q^2 - d)\mu}{q^2} = \mu \qquad (2.56)$$

and

$$\mathbb{V}[x] = \left(2q^2 \cdot \sum_{l=1}^{d} \Gamma_{.l}\Gamma_{.l}'\right) \cdot \frac{1}{2q^2} = \sum_{l=1}^{d} \Gamma_{.l}\Gamma_{.l}' = \Sigma. \qquad (2.57)$$

Looking at the case $d = 2$ helps to understand the choice of the abscissae and weights. It follows that

$$\begin{aligned}
\mathbb{E}[x] &= (\mu - q\Gamma_{.1}) \cdot \frac{1}{2q^2} + (\mu + q\Gamma_{.1}) \cdot \frac{1}{2q^2} \\
&\quad + (\mu - q\Gamma_{.2}) \cdot \frac{1}{2q^2} + (\mu + q\Gamma_{.2}) \cdot \frac{1}{2q^2} \\
&\quad + \mu \frac{q^2 - d}{q^2} \\
&= \frac{\mu q^2 - d\mu + 2\mu}{q^2} = \frac{\mu q^2 - 2\mu + 2\mu}{q^2} = \mu
\end{aligned} \qquad (2.58)$$

and

$$\begin{aligned}
\mathbb{V}[x] &= q^2 \Gamma_{.1}\Gamma_{.1}' \frac{1}{2q^2} + q^2 \Gamma_{.1}\Gamma_{.1}' \frac{1}{2q^2} + q^2 \Gamma_{.2}\Gamma_{.2}' \frac{1}{2p^2} + q^2 \Gamma_{.2}\Gamma_{.2}' \frac{1}{2q^2} \\
&= \Gamma_{.1}\Gamma_{.1}' + \Gamma_{.2}\Gamma_{.2}' = \Sigma.
\end{aligned} \qquad (2.59)$$

The unscented transform of Julier and Uhlman can be used to approximate the moments which have to be calculated within the nonlinear Kalman filter. The resulting unscented Kalman filter algorithm, given by the following pseudocode, has received considerable attention by practitioners and can be considered as a standard filtering algorithm today. Nevertheless, methods of deterministic numerical integration far more exact than the unscented transform exist, which will be described in the third chapter.

Algorithm 3 The unscented Kalman filter algorithm

1: **procedure** UKF
2:    *Definitions*:
3:    $d = \text{dimension}; n = 2d+1$
4:    $\Gamma = \text{chol}(\Sigma)$, lower
5:    $\chi(\mu, \Sigma) = \mu + \text{sgn}(l)\sqrt{d+\kappa}\,\Gamma_{.|l|}$, $l = -d, \ldots, d$
6:    $a_0 = \frac{\kappa}{(d+\kappa)}$
7:    $a_l = \frac{1}{(2(d+\kappa))} = a_{-l}$, $l = 1, \ldots, d$
8:    ───────────────────────────
9:    *Initialization*:
10:    $\mu_{y,1|0};\ \Sigma_{yy,1|0}$
11:    ───────────────────────────
12:    **for** $t=1$ **do**
13:       *Measurement update*
14:    **end for**
15:    **for** $t = 2 : T$ **do**
16:       *Time update*:
17:       $\chi_l = \chi\left(\mu_{y,t-1|t-1}, \Sigma_{yy,t-1|t-1}\right)$
18:       $Y_l = f(\chi_l, x_t, \phi)$, $l = 1, \ldots, n$
19:       $\mu_{y,t|t-1} = \sum_{l=1}^{n} Y_l \alpha_l$
20:       $Y_l^C = Y_l - \mu_{y,t|t-1}$, $l = 1, \ldots, n$
21:       $\Sigma_{yy,t|t-1} = \sum_{l=1}^{n} Y_l^C Y_l^{C'} \alpha_l + \sum_{l=1}^{n} \Omega(\chi_l, x_t, \phi)\alpha_l$
22:       *Measurement update*:
23:       $\chi_l = \chi\left(\mu_{y,t|t-1}, \Sigma_{yy,t|t-1}\right)$
24:       $\chi_l^C = \chi_l - \mu_{y,t|t-1}$, $l = 1, \ldots, n$
25:       $Z_l = h(\chi_l, x_t, \phi)$, $l = 1, \ldots, n$
26:       $\mu_{z,t|t-1} = \sum_{l=1}^{n} Z_l \alpha_l$
27:       $Z_l^C = Z_l - \mu_{z,t|t-1}$, $l = 1, \ldots, n$
28:       $\Sigma_{zz,t|t-1} = \sum_{l=1}^{n} Z_l^C Z_l^{C'} \alpha_l + R(x_t, \phi)$
29:       $\Sigma_{yz,t|t-1} = \sum_{l=1}^{2n+1} \chi_l^C Z_l^{C'} \alpha_l$
30:       $K_t = \Sigma_{yz,t|t-1} \Sigma_{zz,t|t-1}^{-1}$
31:       $\mu_{y,t|t} = \mu_{y,t|t-1} + K_t\left(z_t - \mu_{z,t|t-1}\right)$
32:       $\Sigma_{yy,t|t} = \Sigma_{yy,t|t-1} - K_t \Sigma_{zz,t|t-1} K_t'$
33:    **end for**
34: **end procedure**

2.4 Parameter Estimation

The classic way of parameter estimation is the application of the maximum likelihood method (cf. Tanizaki 2004, pp. 43–46) which can be performed very conveniently using the Kalman filter. Unfortunately, a real-time estimation of the parameters which is, for example, often required in the field of vehicle navigation is usually not possible with this method. The reason for this is that the real-time maximum likelihood estimation requires the optimization of the likelihood function in each time step. Especially in the case of small time intervals and long time series, this is hardly feasible. The described problem can be countered by the use of the Bayesian estimation theory. In this approach, the parameter is treated as a random variable and included in the state vector.

2.4.1 Maximum Likelihood Estimation

The general likelihood function of the measurements Z^t, conditional on the parameter-vector $\boldsymbol{\phi}$, reads

$$L\left(\mathbf{Z}^t|\boldsymbol{\phi}\right) = p\left(z_t, z_{t-1}, \ldots, z_1|\boldsymbol{\phi}\right). \tag{2.60}$$

Using Bayes' theorem and omitting the dependence on $\boldsymbol{\phi}$ for a better readability, it can be decomposed as follows:

$$\begin{aligned} p\left(z_t, z_{t-1}, \ldots, z_1\right) &= p\left(z_t|\mathbf{Z}^{t-1}\right) p\left(\mathbf{Z}^{t-1}\right) \\ &= p\left(z_t|\mathbf{Z}^{t-1}\right) p\left(z_{t-1}|\mathbf{Z}^{T-2}\right) p\left(\mathbf{Z}^{T-2}\right) \\ &\vdots \\ &= \prod_{t=2}^{T} p\left(z_t|\mathbf{Z}^{t-1}\right) p\left(z_1\right). \end{aligned} \tag{2.61}$$

In the linear case, the densities $p\left(z_t|\mathbf{Z}^{t-1}\right)$ are Gaussian with the conditional moments $\boldsymbol{\mu}_{z,t|t-1}$ and $\boldsymbol{\Sigma}_{zz,t|t-1}$. Therefore, the likelihood function reads

$$L\left(\mathbf{Z}^t|\boldsymbol{\phi}\right) = \prod_{t=2}^{T} \mathcal{N}\left(z_t; \boldsymbol{\mu}_{z,t|t-1}, \boldsymbol{\Sigma}_{zz,t|t-1}\right) p\left(z_1\right) \tag{2.62}$$

By using the difference $v_t = z_t - \mu_{z,t|t-1}$, which is also known as the *prediction-error* (cf. Singer 1999, p. 86), one arrives at the prediction-error decomposition of the likelihood function (Schweppe 1965):

$$L\left(Z^t|\phi\right) = \prod_{t=2}^{T} \mathcal{N}\left(v_t; 0, \Sigma_{zz,t|t-1}\right) p\left(z_1\right) \tag{2.63}$$

$$= \prod_{t=2}^{T} \frac{1}{\sqrt{2\pi \Sigma_{zz,t|t-1}}} e^{-\frac{1}{2}\operatorname{tr}\left[\Sigma_{zz,t|t-1}^{-1} v_t v_t'\right]} p\left(z_1\right). \tag{2.64}$$

The needed quantities v_t and $\Sigma_{zz,t|t-1}$ are a byproduct of the Kalman filter algorithm and produced in every time step, which makes the evaluation of the likelihood function very easy. Furthermore, due to the sequential manner in which the likelihood evaluation is performed, no huge matrices must be processed. This greatly contributes to the efficiency of the estimation procedure. If the state-space model is nonlinear, the same estimation procedure can be applied. The main difference to the linear case is the fact that the choice of the normal distribution as likelihood function is only an approximation to the real unknown likelihood function. Therefore, the parameter estimates will be biased.

2.4.2 Bayesian Parameter Estimation

Parameters contained in the drift function can be estimated sequentially using the nonlinear Kalman filter by including them in the state vector. As a consequence, those parameters are then treated as random variables. The resulting state-space model is almost always nonlinear and therefore an approximation to the Bayes estimator density

$$p\left(\phi_t|Z^t\right) = \frac{p\left(z_t|Z^{t-1}, \phi_t\right) \int_{-\infty}^{\infty} p\left(\phi_t|\phi_{t-1}\right) p\left(\phi_{t-1}|Z^{t-1}\right) d\phi_{t-1}}{p\left(z_t|Z^{t-1}\right)} \tag{2.65}$$

arises. However, the nonlinear Kalman filter is not capable of estimating parameters of the diffusion function in a Bayesian manner. In the following remarks, the general approach to Bayesian sequential estimation of parameters with particular focus on the problems inherent to the filtering of diffusion parameters will be examined.

Bayesian Estimation of Drift Parameters As an example, the state-space representation of the ordinary AR(2)-Model with measurement error reads

$$\begin{bmatrix} y_t \\ y_{t-1} \end{bmatrix} = \begin{bmatrix} \phi_1 & \phi_2 \\ 1 & 0 \end{bmatrix} \begin{bmatrix} y_{t-1} \\ y_{t-2} \end{bmatrix} + \begin{bmatrix} \theta \epsilon_t \\ 0 \end{bmatrix} \tag{2.66}$$

$$z_t = y_t + \delta_t.$$

2.4 Parameter Estimation

In order to estimate the drift parameter ϕ_1 in a Bayesian manner, the state vector has to be extended and the resulting nonlinear state-space model reads

$$\begin{bmatrix} y_t \\ y_{t-1} \\ \phi_{1,t} \end{bmatrix} = \begin{bmatrix} y_{t-1}\phi_{1,t-1} + y_{t-2}\phi_2 \\ y_{t-1} \\ \phi_{1,t-1} \end{bmatrix} + \begin{bmatrix} \theta \epsilon_t \\ 0 \\ 0 \end{bmatrix} \qquad (2.67)$$

$$z_t = y_t + \delta_t.$$

The parameter is therefore modeled as a constant, without an additional diffusion term. By the inclusion of the parameter in the state vector, nonlinearities are generated. This can be seen, in this example, by examination of first line of drift matrix, in which products of the state variables can be found. The linkages within the drift term generate correlations between the latent parameter-state and the measurement function. These correlations ensure that the measurements are informative with regard to the update process of the parameter-state. Figure 2.1 shows the filter result of the unscented Kalman filter ($\kappa = 0$) for the extended AR(2)-Model (2.67) with parameters $\phi_1 = 0.5$, $\phi_2 = -0.3$, $\theta = 2$, $R = 0.5$ and initial a priori moments

$$\boldsymbol{\mu}_{y,1|0} = \begin{bmatrix} 0 \\ 0 \\ 0 \end{bmatrix}, \boldsymbol{\Sigma}_{y,1|0} = \begin{bmatrix} 1 & 0 & 0 \\ 0 & 1 & 0 \\ 0 & 0 & 1 \end{bmatrix}.$$

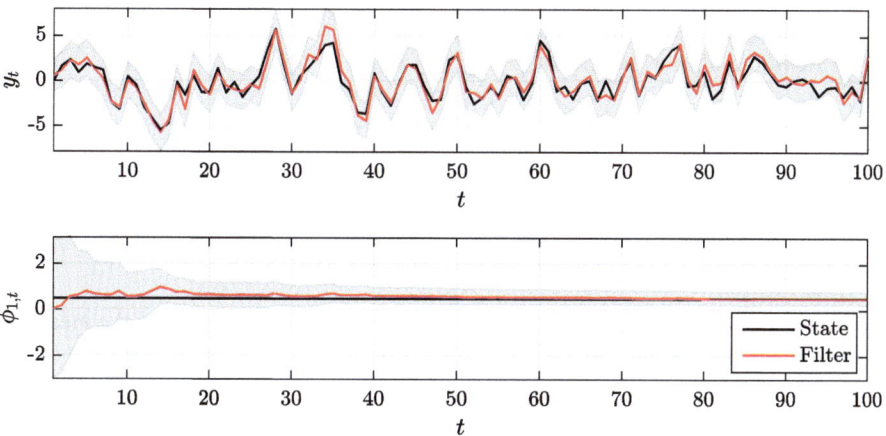

Fig. 2.1 Bayesian parameter estimation of a drift parameter by the example of an AR(2)-Model

The gray area represents the 99% region of highest posterior density (HPD).[6] After few iterations of the filter algorithm, the parameter-state starts converging to the true value of the parameter ($\phi_1 = 0.5$). The estimation of parameters by the extension of the state vector has the major advantage that the estimation process takes place in real-time, and that new information are immediately included in the estimate. A disadvantage is the problem of resulting numerical instabilities, which can be caused by the increase of the dimension the state vector and the resulting nonlinearities.

On the Inability of the Nonlinear Kalman Filter to Filter Diffusion Parameters
Due to the special structure of the equations given by the theorem on normal correlation (2.3), parameters which are part of the diffusion term cannot be filtered by the nonlinear Kalman filter. As can be seen in (2.13) and (2.14), the update of the state depends on the correlation between state and measurement. In order to illustrate the problem, a two-dimensional model with the associated state vector

$$\mathbf{y}_t = \begin{bmatrix} y_t \\ \theta_t \end{bmatrix} \tag{2.68}$$

is used as an example. The two-dimensional state-space model in which the dependence on additional parameters and exogenous variables is left out for simplicity reads

$$\begin{bmatrix} y_t \\ \theta_t \end{bmatrix} = \begin{bmatrix} f(y_{t-1}) \\ \theta_{t-1} \end{bmatrix} + \begin{bmatrix} \theta_t \epsilon_t \\ 0 \end{bmatrix} \tag{2.69}$$

$$z_t = h(y_t) + \delta_t,$$

with the initial a priori moments

$$\boldsymbol{\mu}_{y,1|0} = \begin{bmatrix} \mu_{y,1|0} \\ \mu_{\theta,1|0} \end{bmatrix}$$

$$\boldsymbol{\Sigma}_{yy,1|0} = \begin{bmatrix} \sigma^2_{yy,1|0} & 0 \\ 0 & \sigma^2_{\theta\theta,1|0} \end{bmatrix}. \tag{2.70}$$

To clarify how the independences between the parameter states and the measurements affect the update process, the correlation between the diffusion state θ_t and the measurement z_t will now be calculated. In order to simplify the notation, the

---

[6]The HPD region is constructed as

$$\text{HPD}_{\text{up/down}} = \mu_{y_1,t|t} \pm 3\sqrt{\mathbb{V}(y_{1,t}|\mathbf{Z}^t)}.$$

2.4 Parameter Estimation

expressions $\theta_t - \mu_{\theta,t|t-1}, h(y_t) - \mu_{z,t|t-1}, f(y_{t-1}) - \mu_{y,t-1|t-1}$ and $\theta_{t-1} - \mu_{\theta,t-1|t-1}$ are substituted by $\theta_t^C, h(y_t)^C, f(y_{t-1})^C$ and θ_{t-1}^C.

A crucial quantity for the calculation of the Kalman gain for the state θ_t is the covariance $\text{Cov}\left[\theta_t, z_t | \mathbf{Z}^{t-1}\right]$. Inserting the measurement function yields

$$\text{Cov}\left[\theta_t, z_t | \mathbf{Z}^{t-1}\right] = \text{Cov}\left[\theta_t, h(y_t) + \delta_t | \mathbf{Z}^{t-1}\right]$$
$$= \text{Cov}\left[\theta_t, h(y_t) | \mathbf{Z}^{t-1}\right] + \underbrace{\text{Cov}\left[\theta_t, \delta_t | \mathbf{Z}^{t-1}\right]}_{=0} \quad (2.71)$$
$$= \text{Cov}\left[\theta_t, h(y_t) | \mathbf{Z}^{t-1}\right].$$

The variables y_t and θ_t are uncorrelated (2.70). As with respect to Gaussian random variables uncorrelatedness implies independence, the normal density employed in the integral representation,

$$\text{Cov}\left[\theta_t, z_t | \mathbf{Z}^{t-1}\right] = \int_{-\infty}^{\infty} \int_{-\infty}^{\infty} \theta_t^C h(y_t)^C \mathcal{N}\left(y_t; \mu_{y,t|t-1}, \Sigma_{yy,t|t-1}\right) d\theta_t dy_t, \quad (2.72)$$

can be factorized. This leads to:

$$\text{Cov}\left[\theta_t, z_t | \mathbf{Z}^{t-1}\right] = \underbrace{\int_{-\infty}^{\infty} \theta_t^C \mathcal{N}\left(\theta_t; \mu_{\theta,t|t-1}, \sigma^2_{\theta\theta,t|t-1}\right) d\theta_t}_{=0}$$
$$\cdot \int_{-\infty}^{\infty} h(y_t)^C \mathcal{N}\left(y_t; \mu_{y,t|t-1}, \sigma^2_{yy,t|t-1}\right) dy_t \quad (2.73)$$
$$= 0.$$

Applying this to the Kalman gain,

$$\mathbf{K}_t = \begin{bmatrix} \text{Cov}\left[y_t, z_t | \mathbf{Z}^{t-1}\right] \\ \text{Cov}\left[\theta_t, z_t | \mathbf{Z}^{t-1}\right] \end{bmatrix} \mathbb{V}\left[z_t | \mathbf{Z}^{t-1}\right]^{-1}, \quad (2.74)$$

it shows that the entry of $\Sigma_{yz,t|t-1}$ containing the state θ_t is equal to zero:

$$\mathbf{K}_t = \begin{bmatrix} \text{Cov}\left[y_t, z_t | \mathbf{Z}^{t-1}\right] \\ 0 \end{bmatrix} \mathbb{V}\left[z_t | \mathbf{Z}^{t-1}\right]^{-1}. \quad (2.75)$$

As a consequence, $\mathbf{K}_t$ contains no gain for the diffusion parameter-state and it follows:

$$\mathbf{K}_t = \begin{bmatrix} K_{y_t} \\ 0 \end{bmatrix} \quad (2.76)$$
$$\Rightarrow \mu_{\theta,t|t} = \mu_{\theta,t|t-1}.$$

Moreover, the a posteriori variance reads

$$\Sigma_{yy,t|t} = \begin{bmatrix} \sigma_{yy,t|t-1}^2 & 0 \\ 0 & \sigma_{\theta\theta,t|t-1}^2 \end{bmatrix} - \begin{bmatrix} \text{Cov}[y_t, z_t|\mathbf{Z}^{t-1}] \\ \text{Cov}[\theta_t, z_t|\mathbf{Z}^{t-1}] \end{bmatrix} \mathbb{V}[z_t|\mathbf{Z}^{t-1}]^{-1} \begin{bmatrix} \text{Cov}[y_t, z_t|\mathbf{Z}^{t-1}] \\ \text{Cov}[\theta_t, z_t|\mathbf{Z}^{t-1}] \end{bmatrix}'$$

$$= \begin{bmatrix} \sigma_{yy,t|t-1}^2 & 0 \\ 0 & \sigma_{\theta\theta,t|t-1}^2 \end{bmatrix} - \begin{bmatrix} \text{Cov}[y_t, z_t|\mathbf{Z}^{t-1}] \\ 0 \end{bmatrix} \mathbb{V}[z_t|\mathbf{Z}^{t-1}]^{-1} \begin{bmatrix} \text{Cov}[y_t, z_t|\mathbf{Z}^{t-1}] \\ 0 \end{bmatrix}'$$

$$= \begin{bmatrix} \sigma_{yy,t|t-1}^2 & 0 \\ 0 & \sigma_{\theta\theta,t|t-1}^2 \end{bmatrix} - \begin{bmatrix} K_{y_t} \\ 0 \end{bmatrix} \begin{bmatrix} \text{Cov}[y_t, z_t|\mathbf{Z}^{t-1}] \\ 0 \end{bmatrix}' = \begin{bmatrix} \sigma_{yy,t|t}^2 & 0 \\ 0 & \sigma_{\theta\theta,t|t-1}^2 \end{bmatrix}$$

(2.77)

what shows that no covariance between y_t and θ_t is generated within the measurement update. The same situation is present with regard to the time update. The time update of the a posteriori covariance,

$$\begin{aligned} \text{Cov}[y_t, \theta_t|\mathbf{Z}^{t-1}] &= \text{Cov}[f(y_{t-1}) + \theta_t \epsilon_t, \theta_t|\mathbf{Z}^{t-1}] \\ &= \text{Cov}[f(y_{t-1}), \theta_t|\mathbf{Z}^{t-1}] + \text{Cov}[\theta_t \epsilon_t, \theta_t|\mathbf{Z}^{t-1}] \\ &= \text{Cov}[f(y_{t-1}), \theta_t|\mathbf{Z}^{t-1}] \\ &\quad + \underbrace{\mathbb{E}[\theta_t^2 \epsilon_t|\mathbf{Z}^{t-1}]}_{=0} - \underbrace{\mathbb{E}[\theta_t \epsilon_t|\mathbf{Z}^{t-1}]}_{=0} \mathbb{E}[\theta_t|\mathbf{Z}^{t-1}] \\ &= \text{Cov}[f(y_{t-1}), \theta_t|\mathbf{Z}^{t-1}], \end{aligned}$$

(2.78)

is equal to zero as well which can be shown by inspecting its integral representation:

$$\text{Cov}[y_t, \theta_t|\mathbf{Z}^{t-1}] = \int_{-\infty}^{\infty} \int_{-\infty}^{\infty} f(y_{t-1})^C \theta_{t-1}^C \mathcal{N}(y_{t-1}; \mu_{y,t-1|t-1}, \Sigma_{yy,t-1|t-1}).$$

(2.79)

By the same argument as in (2.73), the integral can be factorized, so that

$$\text{Cov}[y_t, \theta_t|\mathbf{Z}^{t-1}] = \int_{-\infty}^{\infty} f(y_{t-1})^C \mathcal{N}(y_{t-1}; \mu_{y,t-1|t-1}, \sigma_{yy,t-1|t-1}^2) dy_{t-1}$$

$$\cdot \underbrace{\int_{-\infty}^{\infty} \theta_{t-1}^C \mathcal{N}(\theta_{t-1}; \mu_{\theta,t-1|t-1}, \sigma_{\theta\theta,t-1|t-1}^2) d\theta_{t-1}}_{=0}$$

(2.80)

$$= 0.$$

2.4 Parameter Estimation

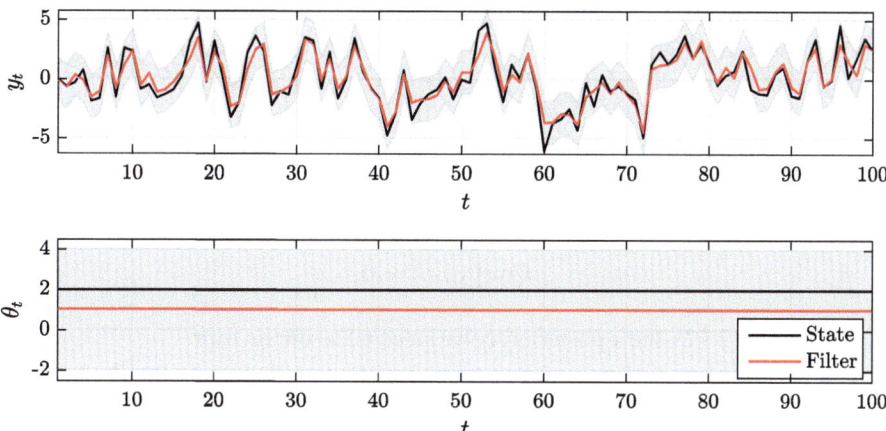

Fig. 2.2 Bayesian parameter estimation of a diffusion parameter without initial covariance by the example of an AR(2)-Model

Summarized, the initial zero-correlation of y_t and θ_t is transported through all time steps. As a result, no update for θ_t is produced at any time. Figure 2.2 shows the filter result of the unscented Kalman filter ($\kappa = 0$) for the extended AR(2)-Model

$$\begin{bmatrix} y_t \\ y_{t-1} \\ \theta_t \end{bmatrix} = \begin{bmatrix} y_{t-1}\phi_1 + y_{t-2}\phi_2 \\ y_{t-1} \\ \theta_{t-1} \end{bmatrix} + \begin{bmatrix} \theta_{t-1}\epsilon_t \\ 0 \\ 0 \end{bmatrix} \quad (2.81)$$

$$z_t = y_t + \delta_t.$$

with parameters $\phi_1 = 0.5$, $\phi_2 = -0.3$, $\theta = 2$, $R = 0.5$ and starting values

$$\boldsymbol{\mu}_{y,1|0} = \begin{bmatrix} 0 \\ 0 \\ 1 \end{bmatrix}, \boldsymbol{\Sigma}_{yy,1|0} = \begin{bmatrix} 1 & 0 & 0 \\ 0 & 1 & 0 \\ 0 & 0 & 1 \end{bmatrix}.$$

As can be seen, the filter trajectory for θ_t remains in its starting point.

Even if the initial covariance is chosen differently from zero,

$$\boldsymbol{\Sigma}_{yy,1|0} = \begin{bmatrix} \sigma^2_{yy,1|0} & \sigma_{y\theta,1|0} \\ \sigma_{\theta y,1|0} & \sigma^2_{\theta\theta,1|0} \end{bmatrix}, \quad (2.82)$$

the nonlinear Kalman filter is not able to filter the state θ_t because over time, the initial covariance between y_t and θ_t drops down to zero. As a consequence, the updating process for θ_t comes to a halt, as already made clear before. To give an analytical explanation for this fact, however, proves to be difficult. First it is going to be shown that the a posteriori covariance $\text{Cov}\left[y_t, \theta_t | \mathbf{Z}^t\right]$ is always smaller than the a priori covariance $\text{Cov}\left[y_t, \theta_t | \mathbf{Z}^{t-1}\right]$ in absolute value. Because y_t and θ_t are supposed to be initially correlated, θ_t can be rewritten in terms of y_t:

$$\theta_t = \alpha_t + \beta_t y_t + \epsilon_t, \tag{2.83}$$

with

$$\text{Cov}\left[y_t, \epsilon_t\right] = 0. \tag{2.84}$$

According to (2.77), the a posteriori covariance takes on the form

$$\text{Cov}\left[y_t, \theta_t | \mathbf{Z}^t\right] = \text{Cov}\left[y_t, \theta_t | \mathbf{Z}^{t-1}\right] - \frac{\text{Cov}\left[y_t, z_t | \mathbf{Z}^{t-1}\right] \text{Cov}\left[\theta_t, z_t | \mathbf{Z}^{t-1}\right]}{\mathbb{V}\left[z_t | \mathbf{Z}^{t-1}\right]} \tag{2.85}$$

Using (2.83) it turns out that

$$\begin{aligned}\text{Cov}\left[\theta_t, z_t | \mathbf{Z}^{t-1}\right] &= \text{Cov}\left[\alpha_t + \beta_t y_t + \epsilon_t, z_t | \mathbf{Z}^{t-1}\right] \\ &= \beta_t \text{Cov}\left[y_t, z_t | \mathbf{Z}^{t-1}\right],\end{aligned} \tag{2.86}$$

with β_t as the usual regression coefficient,

$$\beta_t = \frac{\text{Cov}\left[y_t, \theta_t | \mathbf{Z}^{t-1}\right]}{\mathbb{V}\left[y_t | \mathbf{Z}^{t-1}\right]}. \tag{2.87}$$

It follows that the a posteriori covariance can be rewritten into

$$\text{Cov}\left[y_t, \theta_t | \mathbf{Z}^t\right] = \text{Cov}\left[y_t, \theta_t | \mathbf{Z}^{t-1}\right] - \frac{\text{Cov}\left[y_t, z_t | \mathbf{Z}^{t-1}\right] \beta_t \text{Cov}\left[y_t, z_t | \mathbf{Z}^{t-1}\right]}{\mathbb{V}\left[z_t | \mathbf{Z}^{t-1}\right]}. \tag{2.88}$$

After inserting expression (2.87), $\text{Cov}\left[y_t, \theta_t | \mathbf{Z}^{t-1}\right]$ can be rearranged to

$$\text{Cov}\left[y_t, \theta_t | \mathbf{Z}^t\right] = \text{Cov}\left[y_t, \theta_t | \mathbf{Z}^{t-1}\right] - \frac{\text{Cov}\left[y_t, z_t | \mathbf{Z}^{t-1}\right] \frac{\text{Cov}\left[y_t, \theta_t | \mathbf{Z}^{t-1}\right]}{\mathbb{V}\left[y_t | \mathbf{Z}^{t-1}\right]} \text{Cov}\left[y_t, z_t | \mathbf{Z}^{t-1}\right]}{\mathbb{V}\left[z_t | \mathbf{Z}^{t-1}\right]} \tag{2.89}$$

$$= \text{Cov}\left[y_t, \theta_t | \mathbf{Z}^{t-1}\right] \cdot \left(1 - \frac{\text{Cov}\left[y_t, z_t | \mathbf{Z}^{t-1}\right]^2}{\mathbb{V}\left[y_t | \mathbf{Z}^{t-1}\right] \mathbb{V}\left[z_t | \mathbf{Z}^{t-1}\right]}\right). \tag{2.90}$$

2.4 Parameter Estimation

By the Cauchy–Schwarz inequality (cf. Florescu 2014, p. 135), it is

$$\text{Cov}\left[y_t, z_t | \mathbf{Z}^{t-1}\right]^2 \leq \mathbb{V}\left[y_t | \mathbf{Z}^{t-1}\right] \mathbb{V}\left[z_t | \mathbf{Z}^{t-1}\right]. \tag{2.91}$$

Therefore it holds that

$$\left|\text{Cov}\left[y_t, \theta_t | \mathbf{Z}^t\right]\right| < \left|\text{Cov}\left[y_t, \theta_t | \mathbf{Z}^{t-1}\right]\right|. \tag{2.92}$$

Thus, the absolute value of the covariance between y_t and θ_t is always reduced in the measurement update and it must be shown that this is also the case in each time update. The time update of the covariance reads

$$\text{Cov}\left[y_t, \theta_t | \mathbf{Z}^{t-1}\right] = \text{Cov}\left[f(y_{t-1}), \theta_{t-1} | \mathbf{Z}^{t-1}\right]. \tag{2.93}$$

From a heuristic point of view it is comprehensible that the linear relationship between two correlated random variables y_{t-1} and θ_{t-1} in many cases will be larger than the relationship between $f(y_{t-1})$ and θ_{t-1}, where f is a nonlinear function which has specific characteristics. By the nonlinear transformation of y_{t-1} the originally present covariance between y_{t-1} and θ_{t-1} decreases in absolute value. As already described, the absolute covariance is always reduced in the measurement update and therefore, the focus will now be placed solely on the time update.

Algorithm 4 Decreasing covariance over time

1: **procedure** DECREASING COV
2:    *Initialization*:
3:    $\boldsymbol{\mu}; \boldsymbol{\Sigma}; [y_{t-1}, \theta_{t-1}] \sim \mathcal{N}(\boldsymbol{\mu}, \boldsymbol{\Sigma})$
4: ———
5:    $\boldsymbol{\mu}_1 = \boldsymbol{\mu}$
6:    $\boldsymbol{\Sigma}_1 = \boldsymbol{\Sigma}$
7:    **for** $t = 1 : p$ **do**
8:       $\boldsymbol{\mu}_t = \mathbb{E}\left[[f(y_{t-1}), \theta_{t-1}] | \mathbf{Z}_{t-1}\right]$
9:       $\boldsymbol{\Sigma}_t = \mathbb{V}\left[[f(y_{t-1}), \theta_{t-1}] | \mathbf{Z}_{t-1}\right]$
10:      $[y_{t-1}, \theta_{t-1}] \sim \mathcal{N}(\boldsymbol{\mu}_t, \boldsymbol{\Sigma}_t)$
11:   **end for**
12: **end procedure**

Algorithm 4 describes an iterative implementation of time updates. Own simulation studies using the algorithm in combination with a variety of state-space models have shown that, over time, the absolute covariance between y_{t-1} and θ_{t-1} becomes smaller and finally approaches zero. Of course, in general this is not true and counterexamples can be found easily. For example, if $f(y_{t-1}) = e^{y_{t-1}}$, the covariance will tend to infinity. The presumption is, however, that the stated interrelation is true with respect to the class of functions which is suitable for the construction of state-space models. In order to give an explanation for this behaviour, analogous to (2.83), θ_{t-1} gets rewritten in terms of y_{t-1}:

$$\theta_{t-1} = \alpha_t + \beta_t y_{t-1} + \epsilon_t. \tag{2.94}$$

The covariance then reads

$$\text{Cov}\left[f(y_{t-1}), \theta_{t-1} | \mathbf{Z}^{t-1}\right] = \beta_t \text{Cov}\left[f(y_{t-1}), y_{t-1} | \mathbf{Z}^{t-1}\right], \tag{2.95}$$

with

$$\beta_t = \frac{\text{Cov}\left[\theta_{t-1}, y_{t-1} | \mathbf{Z}^{t-1}\right]}{\mathbb{V}\left[y_{t-1} | \mathbf{Z}^{t-1}\right]}. \tag{2.96}$$

This indicates that $\text{Cov}\left[f(y_{t-1}), \theta_{t-1} | \mathbf{Z}^{t-1}\right]$ can be written in a recursive manner, as

$$\begin{aligned}
\text{Cov}\left[f(y_{t-1}), \theta_{t-1} | \mathbf{Z}^{t-1}\right] &= \frac{\text{Cov}\left[y_{t-1}, \theta_{t-1} | \mathbf{Z}^{t-1}\right]}{\mathbb{V}\left[y_{t-1} | \mathbf{Z}^{t-1}\right]} \text{Cov}\left[f(y_{t-1}), y_{t-1} | \mathbf{Z}^{t-1}\right] \\
&= \frac{\text{Cov}\left[f(y_{t-1}), y_{t-1} | \mathbf{Z}^{t-1}\right]}{\mathbb{V}\left[y_{t-1} | \mathbf{Z}^{t-1}\right]} \text{Cov}\left[y_{t-1}, \theta_{t-1} | \mathbf{Z}^{t-1}\right] \\
&= \tilde{\beta}_t \text{Cov}\left[y_{t-1}, \theta_{t-1} | \mathbf{Z}^{t-1}\right].
\end{aligned} \tag{2.97}$$

The parameter $\tilde{\beta}_t$ is the regression parameter for the regression of $f(y_{t-1})$ on y_{t-1}. To guarantee that $\text{Cov}\left[f(y_{t-1}), \theta_{t-1} | \mathbf{Z}^{t-1}\right]$ decreases to zero, it must be that $\left|\tilde{\beta}_t\right| < 1$ from a certain t on. This proves to be correct with respect to the conducted simulation studies and therefore represents the reason for the fact that the diffusion parameter θ_t cannot be filtered by the nonlinear Kalman filter, even if a initial covariance between y_1 and θ_1 is given. The exact conditions which the drift function f has to fulfill so that $\left|\tilde{\beta}_t\right| < 1$ have not become completely clear. However, it seems very likely that the behaviour of the slope of the function plays a crucial role.

Figure 2.3 shows the filter solution of the unscented Kalman filter ($\kappa = 0$) for the model (2.81), this time with initial covariance

$$\Sigma_{yy,1|0} = \begin{bmatrix} 1 & 0.1 & 0.1 \\ 0.1 & 1 & 0.1 \\ 0.1 & 0.1 & 1 \end{bmatrix}.$$

2.5 Conditional Filtering

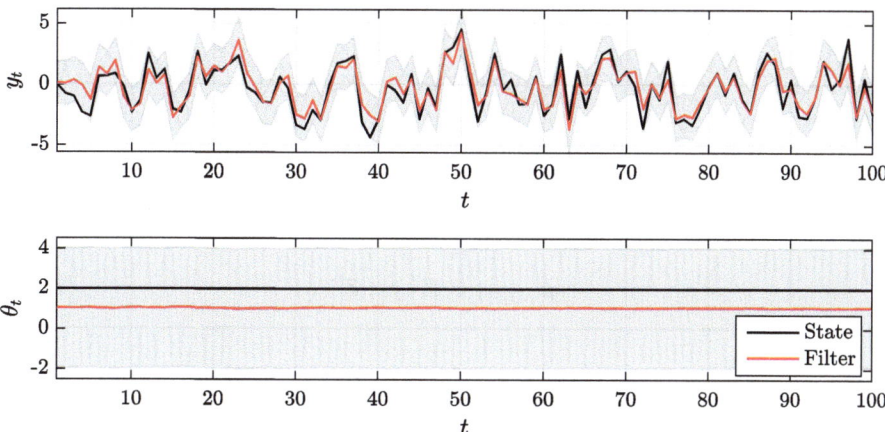

Fig. 2.3 Bayesian parameter estimation of a diffusion parameter with initial covariance by the example of an AR(2)-Model

At the beginning of the filtering process, some activity in the filter trajectory for θ_t can be observed but convergence to the value of the diffusion parameter ($\theta = 2$) does not take place. As time proceeds, the activity is becoming weaker and finally drops down to almost zero.

2.5 Conditional Filtering

An extension to the nonlinear Kalman filter which makes the filtering of diffusion parameters possible has been proposed by Singer (2015). Instead of approximating the filter densities $p\left(y_t | Z^{t-i}\right)$ ($i = 0, 1$) by the Gaussian densities $\mathcal{N}\left(\mu_{y,t|t-i}, \Sigma_{yy,t|t-i}\right)$, the state vector y_t of the general Gaussian discrete state-space model, (2.1, 2.2), is decomposed into two parts, denoted as $y_{1,t}$ and $y_{2,t}$. Referencing to the state-space model (2.69), to filter the diffusion parameter θ the state $y_{2,t}$ would be chosen to be θ_t. Because the filter can be applied to all sorts of state-space models, instead of restricting the notation to the case of diffusion parameters, the algorithm will be described in a general form.

The basic idea is to factor the unknown density as follows:

$$\begin{aligned} p\left(y_t | Z^{t-i}\right) &= p\left(y_{1,t}, y_{2,t} | Z^{t-i}\right) \\ &= p\left(y_{1,t} | y_{2,t}, Z^{t-i}\right) p\left(y_{2,t} | Z^{t-i}\right) \\ &\approx \mathcal{N}\left(\mu_{y_{1,t}|y_{2,t},Z^{t-i}}, \Sigma_{y_{1,t}y_{1,t}|y_{2,t},Z^{t-i}}\right) \cdot \mathcal{N}\left(\mu_{y_2,t|t-i}, \Sigma_{y_2 y_2,t|t-i}\right), \end{aligned} \tag{2.98}$$

with

$$\mu_{y_{1,t}|y_{2,t},Z^{t-i}} = \mathbb{E}\left[y_{1,t}|y_{2,t},Z^{t-i}\right] \qquad (2.99)$$

$$\Sigma_{y_{1,t}y_{1,t}|y_{2,t},Z^{t-i}} = \mathbb{V}\left[y_{1,t}|y_{2,t},Z^{t-i}\right] \qquad (2.100)$$

$$\mu_{y_{2,t}|t-i} = \mathbb{E}\left[y_{2,t}|Z^{t-i}\right] \qquad (2.101)$$

$$\Sigma_{y_2 y_2,t|t-i} = \mathbb{V}\left[y_{2,t}|Z^{t-i}\right]. \qquad (2.102)$$

The result is a new state-space model in which the functions f and g are decomposed into two sub-functions:

$$y_{1,t} = f_1\left(y_{1,t-1}, y_{2,t-1}, x_t, \phi\right) + g_1\left(y_{1,t-1}, y_{2,t-1}, x_t, \phi\right) \epsilon_t \qquad (2.103)$$
$$y_{2,t} = f_2\left(y_{1,t-1}, y_{2,t-1}, x_t, \phi\right) + g_2\left(y_{1,t-1}, y_{2,t-1}, x_t, \phi\right) \epsilon_t \qquad (2.104)$$
$$z_t = h\left(y_{1,t}, y_{2,t}, x_t, \phi\right) + \delta_t. \qquad (2.105)$$

Dropping the function arguments, the equations for the time updates read

$$\begin{aligned}\mu_{y_{1,t}|y_{2,t},Z^{t-1}} &= \mathbb{E}\left[f_1 + g_1\epsilon_t|y_{2,t},Z^{t-1}\right] \\ &= \mathbb{E}\left[f_1|y_{2,t},Z^{t-1}\right]\end{aligned} \qquad (2.106)$$

$$\begin{aligned}\Sigma_{y_{1,t}y_{1,t}|y_{2,t},Z^{t-1}} &= \mathbb{V}\left[f_1|y_{2,t},Z^{t-1}\right] + \mathbb{V}\left[g_1\epsilon_t|y_{2,t},Z^{t-1}\right] \\ &= \mathbb{V}\left[f_1|y_{2,t},Z^{t-1}\right] + \mathbb{E}\left[g_1 g_1'|y_{2,t},Z^{t-1}\right]\mathbb{V}\left[\epsilon_t\right] \\ &= \mathbb{V}\left[f_1|y_{2,t},Z^{t-1}\right] + \mathbb{E}\left[\Omega_1|y_{2,t},Z^{t-1}\right]\end{aligned} \qquad (2.107)$$

$$\begin{aligned}\mu_{y_{2,t}|t-1} &= \mathbb{E}\left[f_2 + g_2\epsilon_t|Z^{t-1}\right] \\ &= \mathbb{E}\left[f_2|Z^{t-1}\right]\end{aligned} \qquad (2.108)$$

$$\begin{aligned}\Sigma_{y_2 y_2,t|t-1} &= \mathbb{V}\left[f_2|Z^{t-1}\right] + \mathbb{V}\left[g_2\epsilon_t|Z^{t-1}\right] \\ &= \mathbb{V}\left[f_2|Z^{t-1}\right] + \mathbb{E}\left[g_2 g_2'|Z^{t-1}\right]\mathbb{V}\left[\epsilon_t\right] \\ &= \mathbb{V}\left[f_2|Z^{t-1}\right] + \mathbb{E}\left[\Omega_2|Z^{t-1}\right]\end{aligned} \qquad (2.109)$$

2.5 Conditional Filtering

First, the integral representations for the time updates (2.106) and (2.107) will now be derived. To perform these updates, the Gaussian density

$$\mathcal{N}\left(\mu_{y_{1,t-1}|y_{2,t},z^{t-1}}, \Sigma_{y_{1,t-1}y_{1,t-1}|y_{2,t},z^{t-1}}\right) \tag{2.110}$$

is required. For simplification reasons it is now assumed that

$$\mathcal{N}\left(\mu_{y_{1,t-1}|y_{2,t},z^{t-1}}, \Sigma_{y_{1,t-1}y_{1,t-1}|y_{2,t},z^{t-1}}\right)$$
$$\approx \mathcal{N}\left(\mu_{y_{1,t-1}|y_{2,t-1},z^{t-1}}, \Sigma_{y_{1,t-1}y_{1,t-1}|y_{2,t-1},z^{t-1}}\right). \tag{2.111}$$

Therefore, after dropping the arguments x_t and ϕ for a better readability, the expressions for the time update of the moments $\mu_{y_{1,t}|y_{2,t},z^t}$ and $\Sigma_{y_{1,t}y_{1,t}|y_{2,t},z^t}$ read

$$\mu_{y_{1,t}|y_{2,t},z^{t-1}} = \int_{-\infty}^{\infty} f_1(y_{1,t-1}, y_{2,t-1}) \mathcal{N}_1 dy_{1,t-1} \tag{2.112}$$

$$\Sigma_{y_{1,t}y_{1,t}|y_{2,t},z^{t-1}} = \int_{-\infty}^{\infty} f_1^C(y_{1,t-1}, y_{2,t-1}) f_1^C(y_{1,t-1}, y_{2,t-1})' \mathcal{N}_1 dy_{1,t-1}$$
$$+ \int_{-\infty}^{\infty} \Omega_1(y_{1,t-1}, y_{2,t-1}) \mathcal{N}_1 dy_{1,t-1}, \tag{2.113}$$

with

$$f_1^C(y_{1,t-1}, y_{2,t-1}) = f_1(y_{1,t-1}, y_{2,t-1}) - \mu_{y_{1,t}|y_{2,t},z^{t-1}} \tag{2.114}$$

and

$$\mathcal{N}_1 = \mathcal{N}\left(y_{1,t-1}; \mu_{y_{1,t-1}|y_{2,t-1},z^{t-1}}, \Sigma_{y_{1,t-1}y_{1,t-1}|y_{2,t-1},z^{t-1}}\right). \tag{2.115}$$

As one possible method of numerical integration, the unscented transform (Sect. 2.3.2) can be used in order to evaluate the moments. For each abscissa $\chi_{2,m}$,

$$\chi_{2,m} = \chi\left(\mu_{y_2,t-1|t-1}, \Sigma_{y_2 y_2,t-1|t-1}\right), m = 1, \ldots, M, \tag{2.116}$$

there exists on set of moments $\mu_{y_{1,t}|y_{2,t}=\chi_{2,m},z^t}$ and $\Sigma_{y_{1,t}y_{1,t}|y_{2,t}=\chi_{2,m},z^t}$. This is because the conditional density $\mathcal{N}\left(\mu_{y_{1,t-1}|y_{2,t-1},z^{t-1}}, \Sigma_{y_{1,t-1}y_{1,t-1}|y_{2,t-1},z^{t-1}}\right)$ is discretized by the introduction of the abscissae $\chi_{2,m}$ with respect to $y_{2,t}$. Consequently, the conditional moments have to be evaluated at the given locations. Thus, for every abscissa $\chi_{2,m}$ one set of abscissae

$$\chi_{1,lm} = \chi\left(\mu_{y_{1,t-1}|y_{2,t-1}=\chi_{2,m},z^{t-1}}, \Sigma_{y_{1,t-1}y_{1,t-1}|y_{2,t-1}=\chi_{2,m},z^{t-1}}\right), l = 1,\ldots,L, \quad (2.117)$$

has to be constructed. The time update of the sets $\left\{\mu_{y_{1,t}|y_{2,t}=\chi_{2,m},z^t}, \Sigma_{y_{1,t}y_{1,t}|y_{2,t}=\chi_{2,m},z^t}\right\}$, $m = 1,\ldots,M$, can then be approximated by the sums

$$\mu_{y_{1,t}|y_{2,t}=\chi_{2,m},z^{t-1}} \approx \sum_{l=1}^{L} f_1\left(\chi_{1,lm}, \chi_{2,m}\right) \alpha_{1,l} \quad (2.118)$$

$$\Sigma_{y_{1,t}y_{1,t}|y_{2,t}=\chi_{2,m},z^{t-1}} \approx \sum_{l=1}^{L} f_1^C\left(\chi_{1,lm}, \chi_{2,m}\right) f_1^C\left(\chi_{1,lm}, \chi_{2,m}\right)' \alpha_{1,l}$$

$$+ \sum_{l=1}^{L} \Omega_1\left(\chi_{1,lm}, \chi_{2,m}\right) \alpha_{1,l}. \quad (2.119)$$

The integral representations for the remaining time updates (2.108) and (2.109) read

$$\mu_{y_2,t|t-1} = \int_{-\infty}^{\infty}\int_{-\infty}^{\infty} f_2\left(y_{1,t-1}, y_{2,t-1}\right) \mathcal{N}_1 \mathcal{N}_2 dy_{1,t-1} dy_{2,t-1} \quad (2.120)$$

$$\Sigma_{y_2 y_2,t|t-1} = \int_{-\infty}^{\infty}\int_{-\infty}^{\infty} f_2^C(y_{1,t-1}, y_{2,t-1}) f_2^C(y_{1,t-1}, y_{2,t-1})' \mathcal{N}_1 \mathcal{N}_2 dy_{1,t-1} dy_{2,t-1}$$

$$+ \int_{-\infty}^{\infty}\int_{-\infty}^{\infty} \Omega_2\left(y_{1,t-1}, y_{2,t-1}\right) \mathcal{N}_1 \mathcal{N}_2 dy_{1,t-1} dy_{2,t-1}, \quad (2.121)$$

with

$$f_2^C\left(y_{1,t-1}, y_{2,t-1}\right) = f_2\left(y_{1,t-1}, y_{2,t-1}\right) - \mu_{y_2,t|t-1} \quad (2.122)$$

and

$$\mathcal{N}_2 = \mathcal{N}\left(y_{2,t-1}; \mu_{y_2,t-1|t-1}, \Sigma_{y_2 y_2,t-1|t-1}\right). \quad (2.123)$$

2.5 Conditional Filtering

Again, the values of the integrals can be approximated using the unscented transform or some other method of numerical integration:

$$\mu_{y_2,t|t-1} \approx \sum_{l=1}^{L} \sum_{m=1}^{M} f_2\left(\chi_{1,lm}, \chi_{2,m}\right) \alpha_{1,l} \alpha_{2,m} \tag{2.124}$$

$$\Sigma_{y_2 y_2,t|t-1} \approx \sum_{l=1}^{L} \sum_{m=1}^{M} f_2^C\left(\chi_{1,lm}, \chi_{2,m}\right) f_2^C\left(\chi_{1,lm}, \chi_{2,m}\right)' \alpha_{1,l} \alpha_{2,m}$$
$$+ \sum_{l=1}^{L} \sum_{m=1}^{M} \Omega_2\left(\chi_{1,lm}, \chi_{2,m}\right) \alpha_{1,l} \alpha_{2,m} \tag{2.125}$$

For the measurement update, the following density is required:

$$\begin{aligned} p\left(y_{1,t}, y_{2,t}|z_t, Z^{t-1}\right) &= p\left(y_{1,t}|y_{2,t}, z_t, Z^{t-1}\right) p\left(y_{2,t}|z_t, Z^{t-1}\right) \\ &= p\left(y_{1,t}|y_{2,t}, Z^t\right) \frac{p\left(y_{2,t}, z_t, Z^{t-1}\right)}{p\left(z_t, Z^{t-1}\right)} \\ &= p\left(y_{1,t}|y_{2,t}, Z^t\right) \frac{p\left(z_t|y_{2,t}, Z^{t-1}\right) p\left(y_{2,t}, Z^{t-1}\right)}{p\left(z_t, Z^{t-1}\right)} \quad (2.126) \\ &= p\left(y_{1,t}|y_{2,t}, Z^t\right) \frac{p\left(z_t|y_{2,t}, Z^{t-1}\right) p\left(y_{2,t}|Z^{t-1}\right)}{p\left(z_t|Z^{t-1}\right)}. \end{aligned}$$

By using the approximation

$$p\left(y_{1,t}|y_{2,t}, Z^t\right) \approx \mathcal{N}\left(y_{1,t}; \mu_{y_{1,t}|y_{2,t},Z^t}, \Sigma_{y_{1,t}y_{1,t}|y_{2,t},Z^t}\right). \tag{2.127}$$

the measurement update of the moments $\mu_{y_{1,t}|y_{2,t},Z^{t-1}}$ and $\Sigma_{y_{1,t}y_{1,t}|y_{2,t},Z^{t-1}}$ can be performed by using the theorem on normal correlation (2.13, 2.14). Setting

$$\mu_{z_t|y_{2,t},Z^{t-1}} = \mathbb{E}\left[z_t|y_{2,t}, Z^{t-1}\right] \tag{2.128}$$

$$\Sigma_{z_t z_t|y_{2,t},Z^{t-1}} = \mathbb{V}\left[z_t|y_{2,t}, Z^{t-1}\right] \tag{2.129}$$

$$\Sigma_{y_{1,t} z_t|y_{2,t},Z^{t-1}} = \text{Cov}\left[y_{1,t}, z_t|y_{2,t}, Z^{t-1}\right], \tag{2.130}$$

the update equations read

$$\mu_{y_{1,t}|y_{2,t},Z^t} = \mu_{y_{1,t}|y_{2,t},Z^{t-1}} + \Sigma_{y_{1,t}z_t|y_{2,t},Z^{t-1}} \Sigma_{z_t z_t|y_{2,t},Z^{t-1}}^{-1} \left(z_t - \mu_{z_t|y_{2,t},Z^{t-1}}\right) \quad (2.131)$$

and

$$\Sigma_{y_{1,t}y_{1,t}|y_{2,t},Z^t} = \Sigma_{y_{1,t}y_{1,t}|y_{2,t},Z^{t-1}} - \Sigma_{y_{1,t}z_t|y_{2,t},Z^{t-1}} \Sigma_{z_t z_t|y_{2,t},Z^{t-1}}^{-1} \Sigma'_{y_{1,t}z_t|y_{2,t},Z^{t-1}}. \quad (2.132)$$

As in the time update, the needed integrals can be approximated by using methods of deterministic numerical integration like the unscented transform. Based on the a priori moments which are known from the time update, the abscissae $\chi_{2,m}$,

$$\chi_{2,m} = \chi\left(\mu_{y_2,t|t-1}, \Sigma_{y_2 y_2, t|t-1}\right), \, m = 1, \ldots, M, \quad (2.133)$$

have to be computed. Additionally, like in the time update, for every abscissa $\chi_{2,m}$, one set of abscissae

$$\chi_{1,lm} = \chi\left(\mu_{y_{1,t}|y_{2,t}=\chi_{2,m},Z^{t-1}}, \Sigma_{y_{1,t}y_{1,t}|y_{2,t}=\chi_{2,m},Z^{t-1}}\right), \, l = 1, \ldots, L, \quad (2.134)$$

has to be calculated. The normal correlation updates of the a priori sets

$$\left\{\mu_{y_{1,t}|y_{2,t}=\chi_{2,m},Z^{t-1}}, \Sigma_{y_{1,t}y_{1,t}|y_{2,t}=\chi_{2,m},Z^{t-1}}\right\}, \, m = 1, \ldots, M, \quad (2.135)$$

are then evaluated according to the discretization given by the integration points $\chi_{2,m}$. The additionally required moments for the update read

$$\mu_{z_t|y_{2,t},Z^{t-1}} = \mathbb{E}\left[h|y_{2,t},Z^{t-1}\right], \quad (2.136)$$

$$\Sigma_{z_t z_t|y_{2,t},Z^{t-1}} = \mathbb{V}\left[h|y_{2,t},Z^{t-1}\right] + R, \quad (2.137)$$

and

$$\Sigma_{y_{1,t}z_t|y_{2,t},Z^{t-1}} = \mathbb{E}\left[\left(y_{1,t} - \mu_{z_t|y_{2,t},Z^{t-1}}\right)\left(h - \mu_{z_t|y_{2,t},Z^{t-1}}\right)\right]. \quad (2.138)$$

After the introduction of the abbreviations

$$y_{1,t}^C = y_{1,t} - \mu_{y_{1,t}|y_{2,t},Z^{t-1}} \quad (2.139)$$

$$\chi_{1,lm}^C = \chi_{1,lm} - \mu_{y_{1,t}|y_{2,t}=\chi_{2,m},Z^{t-1}}, \, l = 1, \ldots, L \quad (2.140)$$

2.5 Conditional Filtering

$$h^C(y_{1,t}, y_{2,t}) = h(y_{1,t-1}, y_{2,t-1}) - \mu_{z_t|y_{2,t},Z^{t-1}} \tag{2.141}$$

$$\mathcal{N} = \mathcal{N}\left(y_{1,t}; \mu_{y_{1,t}|y_{2,t},Z^{t-1}}, \Sigma_{y_{1,t}y_{1,t}|y_{2,t},Z^{t-1}}\right), \tag{2.142}$$

the integral representations can be formulated as:

$$\mu_{z_t|y_{2,t},Z^{t-1}} = \int_{-\infty}^{\infty} h(y_{1,t}, y_{2,t}) \mathcal{N} dy_{1,t} \tag{2.143}$$

$$\Sigma_{z_t z_t|y_{2,t},Z^{t-1}} = \int_{-\infty}^{\infty} h^C(y_{1,t}, y_{2,t}) h^C(y_{1,t}, y_{2,t})' \mathcal{N} dy_{1,t} + R \tag{2.144}$$

$$\Sigma_{y_{1,t} z_t|y_{2,t},Z^{t-1}} = \int_{-\infty}^{\infty} y_{1,t}^C h^C(y_{1,t}, y_{2,t})' \mathcal{N} dy_{1,t}. \tag{2.145}$$

By applying the abscissae and weights, the following sums are then used as approximations to the integrals:

$$\mu_{z_t|y_{2,t}=\chi_{2,m},Z^{t-1}} \approx \sum_{l=1}^{L} h(\chi_{1,lm}, \chi_{2,m}) \alpha_{1,l} \tag{2.146}$$

$$\Sigma_{z_t z_t|y_{2,t}=\chi_{2,m},Z^{t-1}} \approx \sum_{l=1}^{L} h^C(\chi_{1,lm}, \chi_{2,m}) h^C(\chi_{1,lm}, \chi_{2,m})' \alpha_{1,l} + R \tag{2.147}$$

$$\Sigma_{y_{1,t} z_t|y_{2,t}=\chi_{2,m},Z^{t-1}} \approx \sum_{l=1}^{L} \chi_{1,lm}^C \cdot h^C(\chi_{1,lm}, \chi_{2,m})' \alpha_{1,l}. \tag{2.148}$$

After performing the M normal correlation updates, one obtains the a posteriori sets of moments

$$\left\{\mu_{y_{1,t}|y_{2,t}=\chi_{2,m},Z^t}, \Sigma_{y_{1,t}y_{1,t}|y_{2,t}=\chi_{2,m},Z^t}\right\}, \quad m = 1, \ldots, M. \tag{2.149}$$

In order to obtain the expected value and the variance of $y_{1,t}|Z^t$ unconditional on $y_{2,t}$, the conditional moments have to be consolidated. This is achieved by using the law of total expectation and the law of total variance. The calculations proceed as follows:

$$\mu_{y_1,t|t} = \mathbb{E}\left[y_{1,t}|Z^t\right] = \mathbb{E}\left[\mathbb{E}\left[y_{1,t}|y_{2,t}, Z^t\right]\bigg|Z^t\right] \approx \sum_{m=1}^{M} \mu_{y_{1,t}|y_{2,t}=\chi_{2,m},Z^{t-1}} \cdot \alpha_{2,m}, \tag{2.150}$$

$$\Sigma_{y_1 y_1, t|t} = \mathbb{V}[y_{1,t}|Z^t] = \mathbb{E}\left[\mathbb{V}[y_{1,t}|y_{2,t}, Z^t]\Big|Z^t\right] + \mathbb{V}\left[\mathbb{E}[y_{1,t}|y_{2,t}, Z^t]\Big|Z^t\right]$$

$$\approx \sum_{m=1}^{M} \Sigma_{y_{1,t} y_{1,t}|y_{2,t}=\chi_{2,m}, Z^t} \cdot \alpha_{2,m}$$

$$+ \sum_{m=1}^{M} \mu_{y_{1,t}|y_{2,t}=\chi_{2,m}, Z^t}^{C} \cdot \mu_{y_{1,t}|y_{2,t}=\chi_{2,m}, Z^t}^{C'} \cdot \alpha_{2,m}, \quad (2.151)$$

with

$$\mu_{y_{1,t}|y_{2,t}=\chi_{2,m}, Z^t}^{C} = \mu_{y_{1,t}|y_{2,t}=\chi_{2,m}, Z^t} - \mu_{y_1, t|t}, \quad m = 1, \ldots, M. \quad (2.152)$$

As (2.126) shows, the a posteriori distribution of the state $y_{2,t}|Z^t$ reads

$$p(y_{2,t}|Z^t) = \frac{p(z_t|y_{2,t}, Z^{t-1}) p(y_{2,t}|Z^{t-1})}{p(z_t|Z^{t-1})}, \quad (2.153)$$

with

$$p(z_t|y_{2,t}, Z^{t-1}) = \mathcal{N}\left(z_t; \mu_{z_t|y_{2,t}, Z^{t-1}}, \Sigma_{z_t z_t|y_{2,t}, Z^{t-1}}\right). \quad (2.154)$$

Here it becomes apparent, how the influence of the state $y_{2,t}$ is incorporated into the measurement update. Since the moments $\mu_{z_t|y_{2,t}, Z^{t-1}}$ and $\Sigma_{z_t z_t|y_{2,t}, Z^{t-1}}$ are calculated with respect to the density $\mathcal{N}_1$ (2.144, 2.145), they are in general nonlinear functions of $y_{2,t}$. Consequently, the new measurement is informative for the state $y_{2,t}$ (cf. Singer 2015, p. 2479). By applying, for instance, the unscented transform, the likelihood of the data z_t,

$$L(z_t) = p(z_t|Z^{t-1}) = \int_{-\infty}^{\infty} p(z_t|y_{2,t}, Z^{t-1}) p(y_{2,t}|Z^{t-1}) \, dy_{2,t}, \quad (2.155)$$

is approximated by the sum

$$\sum_{m=1}^{M} \mathcal{N}\left(z_t; \mu_{z_t|y_{2,t}=\chi_{2,m}, Z^{t-1}}, \Sigma_{z_t z_t|y_{2,t}=\chi_{2,m}, Z^{t-1}}\right) \alpha_{2,m}. \quad (2.156)$$

Accordingly, the expected value

$$\mu_{y_2|t,t} = \frac{\int_{-\infty}^{\infty} y_{2,t} p(z_t|y_{2,t}, Z^{t-1}) p(y_{2,t}|Z^{t-1}) \, dy_{2,t}}{p(z_t|Z^{t-1})} \quad (2.157)$$

2.5 Conditional Filtering

is approximated by

$$\frac{\sum_{m=1}^{M} \chi_{2,m} \cdot \mathcal{N}\left(z_t; \mu_{z_t|y_{2,t}=\chi_{2,m},Z^{t-1}}, \Sigma_{z_t z_t|y_{2,t}=\chi_{2,m},Z^{t-1}}\right) \alpha_{2,m}}{p\left(z_t|Z^{t-1}\right)}. \quad (2.158)$$

Setting $y_{2,t}^C = y_{2,t} - \mu_{y_2,t|t}$ and $\chi_{2,m}^C = \chi_{2,m} - \mu_{y_2,t|t}$, $m = 1, \ldots, M$, the variance

$$\Sigma_{y_2 y_2|t,t} = \frac{\int_{-\infty}^{\infty} y_{2,t}^C y_{2,t}^{C'} p\left(z_t|y_{2,t}, Z^{t-1}\right) p\left(y_{2,t}|Z^{t-1}\right) dy_{2,t}}{p\left(z_t|Z^{t-1}\right)} \quad (2.159)$$

is approximated by

$$\frac{\sum_{m=1}^{M} \chi_{2,m}^C \chi_{2,m}^{C'} \cdot \mathcal{N}\left(z_t; \mu_{z_t|y_{2,t}=\chi_{2,m},Z^{t-1}}, \Sigma_{z_t z_t|y_{2,t}=\chi_{2,m},Z^{t-1}}\right) \alpha_{2,m}}{p\left(z_t|Z^{t-1}\right)}. \quad (2.160)$$

Now that all the necessary equations have been derived, the conditional filter can be stated in algorithmic form. In the following pseudocode, the unscented transform is exemplary applied for approximation purposes. However, it must be stressed that also any other method of deterministic numerical integration can be used which is suitable for Gaussian integrals.

Algorithm 5 The unscented conditional Kalman filter algorithm

1: **procedure** UCKF
2:     *Definitions*:
3:     $d_1 = \text{dimension}(y_1)$; $d_2 = \text{dimension}(y_2)$; $L = 2d_1 + 1$; $M = 2d_2 + 1$
4:     $\Gamma = \text{chol}(\Sigma)$, lower
5:     $\chi_{1,lm}(\mu, \Sigma) = \mu + \text{sgn}(l) \sqrt{d + \kappa} \Gamma_{.|l|}$, $l = -d_1, \ldots, d_1$
6:     $\chi_{2,m}(\mu, \Sigma) = \mu + \text{sgn}(m) \sqrt{d + \kappa} \Gamma_{.|m|}$, $m = -d_2, \ldots, d_2$
7:     $a_{i,0} = \frac{\kappa}{(d_i + \kappa)}$, $i = 1, 2$
8:     $a_{1,l} = \frac{1}{(2(d_1 + \kappa))} = a_{1,-l}$, $l = 1, \ldots, d_1$
9:     $a_{2,m} = \frac{1}{(2(d_2 + \kappa))} = a_{2,-m}$, $m = 1, \ldots, d_2$
10: ———
11:     *Initialization*:
12:     $\mu_{y_{1,1}|y_{2,1}=\chi_{2,m},Z^0}$; $\Sigma_{y_{1,1}y_{1,1}|y_{2,1}=\chi_{2,m},Z^0}$, $m = 1, 2, \ldots, M$
13:     $\mu_{y_2,1|0}$; $\Sigma_{y_2 y_2,1|0}$
14: ———
15:     **for** $t = 1$ **do**

16:       *Measurement update*
17: **end for**
18: **for** $t = 2 : T$ **do**
19:     *Time update*:
20:     $\chi_{2,m} = \chi_2\left(\mu_{y_2,t-1|t-1}, \Sigma_{y_2 y_2,t-1|t-1}\right)$
21:     **for** $m = 1 : M$ **do**
22:         $\chi_{1,lm} = \chi_1\left(\mu_{y_{1,t-1}|y_{2,t-1}=\chi_{2,m},z^{t-1}}, \Sigma_{y_{1,t-1}y_{1,t-1}|y_{2,t-1}=\chi_{2,m},z^{t-1}}\right)$
23:         $\chi_{lm} = [\chi_{1,lm}, \chi_{2,m}]$
24:         $Y_{1,lm} = f_1(\chi_{lm}, x_t, \phi), \; l = 1, \ldots, L$
25:         $\mu_{y_{1,t}|y_{2,t}=\chi_{2,m},z^{t-1}} = \sum_{l=1}^{L} Y_{1,lm} \alpha_{1,l}$
26:         $Y_{1,lm}^C = Y_{1,lm} - \mu_{y_{1,t}|y_{2,t}=\chi_{2,m},z^{t-1}}, \; l = 1, \ldots, L$
27:         $\Sigma_{y_{1,t}y_{1,t}|y_{2,t}=\chi_{2,m},z^{t-1}} = \sum_{l=1}^{L} Y_{1,lm}^C Y_{1,lm}^{C'} \alpha_{1,l} + \sum_{l=1}^{L} \Omega_1(\chi_{lm}, x_t, \phi) \alpha_{1,l}$
28:     **end for**
29:     $Y_{2,lm} = f_2(\chi_{lm}, x_t, \phi), \; l = 1, \ldots, L, \; m = 1, \ldots, M$
30:     $\mu_{y_2,t|t-1} = \sum_{l=1}^{L} \sum_{m=1}^{M} Y_{2,lm} \alpha_{1,l} \alpha_{2,m}$
31:     $Y_{2,lm}^C = Y_{2,lm} - \mu_{y_2,t|t-1}, \; l = 1, \ldots, L, \; m = 1, \ldots, M$
32:     $\Sigma_{y_2 y_2,t|t-1} = \sum_{l=1}^{L} \sum_{m=1}^{M} Y_{2,lm}^C Y_{2,lm}^{C'} \alpha_{1,l} \alpha_{2,m} + \sum_{l=1}^{L} \sum_{m=1}^{M} \Omega_2(\chi_{lm}, x_t, \phi) \alpha_{1,l} \alpha_{2,m}$
33:     *Measurement update*:
34:     $\chi_{2,m} = \chi_2\left(\mu_{y_2,t|t-1}, \Sigma_{y_2 y_2,t|t-1}\right)$
35:     **for** $m = 1 : M$ **do**
36:         $\chi_{1,lm} = \chi_1\left(\mu_{y_{1,t}|y_{2,t}=\chi_{2,m},z^{t-1}}, \Sigma_{y_{1,t}y_{1,t}|y_{2,t}=\chi_{2,m},z^{t-1}}\right)$
37:         $\chi_{lm} = [\chi_{1,lm}, \chi_{2,m}]$
38:         $Z_l = h(\chi_{lm}, x_t, \phi), \; l = 1, \ldots, L$
39:         $\mu_{z_t|y_{2,t}=\chi_{2,m},z^{t-1}} = \sum_{l=1}^{L} Z_l \alpha_{1,l}$
40:         $Z_l^C = Z_l - \mu_{z_t|y_{2,t}}, \; l = 1, \ldots, L$
41:         $\Sigma_{z_t z_t|y_{2,t}=\chi_{2,m},z^{t-1}} = \sum_{l=1}^{L} Z_l Z_l' \alpha_{1,l} + R(x_t, \phi)$
42:         $\chi_{1,lm}^C = \chi_{1,lm} - \mu_{y_{1,t}|y_{2,t}=\chi_{2,m},z^{t-1}}, \; l = 1, \ldots, L$
43:         $\Sigma_{y_{1,t} z_t|y_{2,t}=\chi_{2,m},z^{t-1}} = \sum_{l=1}^{L} \chi_{1,lm}^C \cdot Z_l^{C'} \alpha_{1,l}$
44:         $K_t = \Sigma_{y_{1,t} z_t|y_{2,t}=\chi_{2,m},z^{t-1}} \Sigma_{z_t z_t|y_{2,t}=\chi_{2,m},z^{t-1}}^{-1}$
45:         $\mu_{y_{1,t}|y_{2,t}=\chi_{2,m},z^t} = \mu_{y_{1,t}|y_{2,t}=\chi_{2,m},z^{t-1}} + K_t\left(z_t - \mu_{z_t|y_{2,t}=\chi_{2,m},z^{t-1}}\right)$
46:         $\Sigma_{y_{1,t}y_{1,t}|y_{2,t}=\chi_{2,m},z^t} = \Sigma_{y_{1,t}y_{1,t}|y_{2,t}=\chi_{2,m},z^{t-1}} - K_t \Sigma_{z_t z_t|y_{2,t}=\chi_{2,m},z^{t-1}}^{-1} K_t'$
47:     **end for**
48:

2.5 Conditional Filtering

49:     *Consolidation for y_1:*

50:     $\mu_{y_1,t|t} = \sum_{m=1}^{M} \mu_{y_1,t|y_{2,t}=\chi_{2,m},Z^{t-1}} \cdot \alpha_{2,m}$

51:     $\mu^C_{y_1,t|y_{2,t}=\chi_{2,m},Z^t} = \mu_{y_1,t|y_{2,t}=\chi_{2,m},Z^t} - \mu_{y_1|t,t}, \quad m = 1, \ldots, M$

52:     $\Sigma_{y_1 y_1,t|t} = \sum_{m=1}^{M} \left(\Sigma_{y_1,t y_1,t|y_{2,t}=\chi_{2,m},Z^t} + \mu^C_{y_1,t|y_{2,t}=\chi_{2,m},Z^t} \cdot \mu^{C'}_{y_1,t|y_{2,t}=\chi_{2,m},Z^t} \right) \cdot \alpha_{2,m}$

53: ───

54:     $L(z_t) = \sum_{m=1}^{M} \mathcal{N}\left(z_t; \mu_{z_t|y_{2,t}=\chi_{2,m},Z^{t-1}}, \Sigma_{z_t z_t|y_{2,t}=\chi_{2,m},Z^{t-1}}\right) \alpha_{2,m}$

55:     $\mu_{y_2,t|t} = \dfrac{\sum_{m=1}^{M} \chi_{2,m} \cdot \mathcal{N}\left(z_t; \mu_{z_t|y_{2,t}=\chi_{2,m},Z^{t-1}}, \Sigma_{z_t z_t|y_{2,t}=\chi_{2,m},Z^{t-1}}\right) \alpha_{2,m}}{L(z_t)}$

56:     $\chi^C_{2,m} = \chi_{2,m} - \mu_{y_2|t,t}, \quad m = 1, \ldots, M$

57:     $\Sigma_{y_2 y_2,t|t} = \dfrac{\sum_{m=1}^{M} \chi^C_{2,m} \chi^{C'}_{2,m} \cdot \mathcal{N}\left(z_t; \mu_{z_t|y_{2,t}=\chi_{2,m},Z^{t-1}}, \Sigma_{z_t z_t|y_{2,t}=\chi_{2,m},Z^{t-1}}\right) \alpha_{2,m}}{L(z_t)}$

58:     **end for**

59: **end procedure**

Figure 2.4 shows the filter solution of the unscented conditional Kalman filter ($\kappa = 0$) for the model (2.81) with the state y_t split into the sub-states

$$y_{1,t} = \begin{bmatrix} y_t \\ y_{t-1} \end{bmatrix} \tag{2.161}$$

$$y_{2,t} = \theta_t$$

and the initial a priori moments

$$\mu_{y_{1,1}|y_{2,1}=\chi_{2,m},Z^0} = \begin{bmatrix} 0 \\ 0 \end{bmatrix}, \quad m = 1, \ldots, M$$

$$\Sigma_{y_{1,1} y_{1,1}|y_{2,1}} = \begin{bmatrix} 1 & 0 \\ 0 & 1 \end{bmatrix}, \quad m = 1, \ldots, M \tag{2.162}$$

$$\mu_{y_2,1|0} = 1$$

$$\Sigma_{y_2 y_2,1|0} = 1.$$

In contrast to the nonlinear Kalman filter, the conditional filter is capable of filtering the diffusion state θ_t. The trajectory converges to the true value of the parameter and delivers a good estimate for the value of θ ($\theta = 2$).

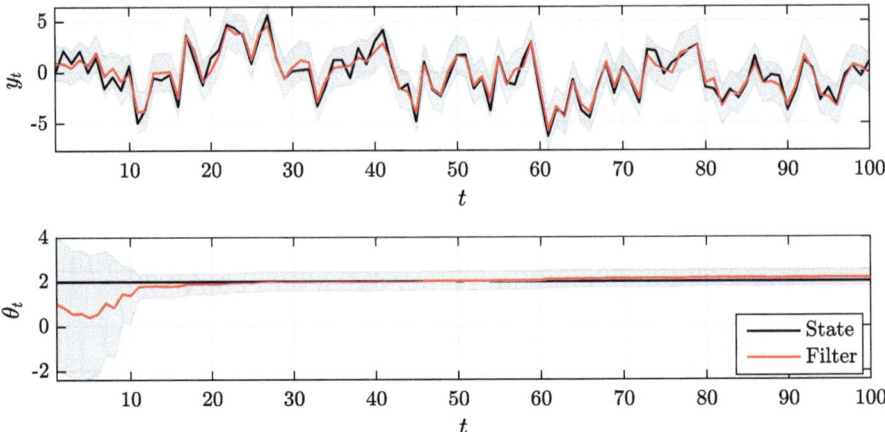

Fig. 2.4 Bayesian parameter estimation of a diffusion parameter by the example of an AR(2)-Model (UCKF)

2.6 Stabilization of Nonlinear Kalman Filter Algorithms

All described nonlinear Kalman filter algorithms have in common that the Cholesky decomposition of the state covariance matrix is needed in the time update and in the measurement update. Due to numerical instabilities, errors like negative variances may occur which lead to the situation that the computed state covariance matrix is not positive definite and therefore no Cholesky decomposition is possible. This must be avoided at all costs, as the filter operation would otherwise be stopped immediately. One solution to this problem is to directly propagate the Cholesky square root of the covariance matrix through the filter iterations instead of the state covariance matrix itself. Algorithms of this kind have been proposed by Potter and Stern (1963), Bellantoni and Dodge (1967), Andrews (1968), van der Merwe and Wan (2001) and others.

In this work, a simple ad-hoc approach will be used, which, despite of its simplicity, delivers good results. In every time update and measurement update first it is checked if one entry of the state covariance matrix is Inf or NaN. In that case, the matrix is set to the identity matrix I. If, however, all entries are real numbers, the eigenvalues of the state covariance matrix are checked. If one eigenvalue is negative or too close to zero, the diagonal elements of the state covariance are first replaced by their absolute values in order to prevent negative variances. From then on, the diagonal elements are increased by a factor until all eigenvalues are bigger than a predefined value.

2.7 Treatment of Missing Data

Algorithm 6 Ensuring the positive definiteness of the state covariance matrix

1: **procedure** SIGMA_POS_DEF
2:    *Initialization*: Σ
3: ───────────────────────────
4:    $\epsilon = 10^{-16}$
5:    **if** one entry of Σ is Inf or NaN **then**
6:       $\Sigma = I$
7:    **end if**
8:    **if** one eigenvalue of Σ is $< \epsilon$ **then**
9:       $D = \text{abs}(\text{diag}(\Sigma))$
10:      **while** one eigenvalue of Σ is $< \epsilon$ **do**
11:         $D = D \cdot 1.01$
12:         $\text{diag}(\Sigma) = D$
13:      **end while**
14:   **end if**
15: **end procedure**

Algorithm 6 shows how the "tweaking" of the state covariance matrix proceeds. Changing the factor 1.01 to a higher value, on the one hand, leads a more rapid achievement of a positive definite state covariance matrix and is therefore time-saving. On the other hand, filter errors tend to increase, because the diagonal elements get enlarged far too much. In order to keep numerical distortions at a minimum the relatively small factor of 1.01 will be maintained.

2.7 Treatment of Missing Data

In practical applications it is often the case that two or more dependent time series have different sampling intervals. Another issue is the possibility of irregularly missing data points because of technical difficulties. The filtering process, whether by the Bayes filter or the Kalman filter, allows a very convenient treatment of missing data. If values of the measurement vector z_t are missing, the measurement update is only performed for the state components for which measurements are available. The remaining components of $\mathbb{E}\left[y_t|Z^t\right]$ are replaced by their time update (prediction).

In the discrete Bayes filter (Algorithm 1), the entries of the Z_l, $l = 1, \ldots, n$, for which the corresponding entries of z_t contain no measurement have to be deleted. Similarly, the corresponding rows and columns of the matrix $R(x_t, \phi)$ must be removed. With respect to the Kalman filter (Algorithms 2, 3 and 5) missing data are treated by deleting the corresponding rows and columns of the Kalman gain K_t

and of the conditional mean $\mathbb{E}\left[z_t|\mathbf{Z}^{t-1}\right]$. For the calculation of the likelihood the vector of prediction-errors $\mathbf{v}_t$ is required and in case of missing measurements, the respective components are set to zero.

The treatment of missing data will now be illustrated by the example of the Ginzburg–Landau model (Ginzburg and Landau 1950), which will be discussed in more detail in Chap. 5. The original model is of deterministic nature and is here extended by random errors and a measurement equation. The resulting stochastic differential equation reads (cf. Singer 2011, p. 20):

$$dy_t = -\left(\alpha y_t + \beta y_t^3\right) dt + g dW_t$$
$$z_{t_i} = y_{t_i} + \delta_{t_i}.$$
(2.163)

The continuous-time process can be observed only at discrete times $\{t_1, t_2, \ldots, t_N\}$ and intermediate times are therefore treated as missing data. In order to apply a discrete filter to the stochastic differential equation, it must be discretized first. Using the Euler method (cf. Kloeden and Platen 1992, 305–307), the discretization of the differential equation takes on the form

$$y_{t+\Delta t} = y_t - \left(\alpha y_t + \beta y_t^3\right) \Delta t + g\sqrt{\Delta t}\epsilon_t,$$
(2.164)

and one arrives at the following state-space model:

$$\begin{bmatrix} y_t \\ y_{t-1} \end{bmatrix} = \begin{bmatrix} y_{t-1} - \left(\alpha y_{t-1} + \beta y_{t-1}^3\right) \Delta t \\ y_{t-1} \end{bmatrix} + \begin{bmatrix} g\sqrt{\Delta t}\epsilon_t \\ 0 \end{bmatrix}$$
$$z_{t_i} = y_{t_i} + \delta_{t_i}.$$
(2.165)

Figure 2.5 shows the filter solution of the unscented Kalman filter ($\kappa = 0$) with the parameter values $\alpha = -1$, $\beta = 1$, $g = 1$ $R = 0.1$, $\Delta t = 0.1$, $T = 100$ (1000

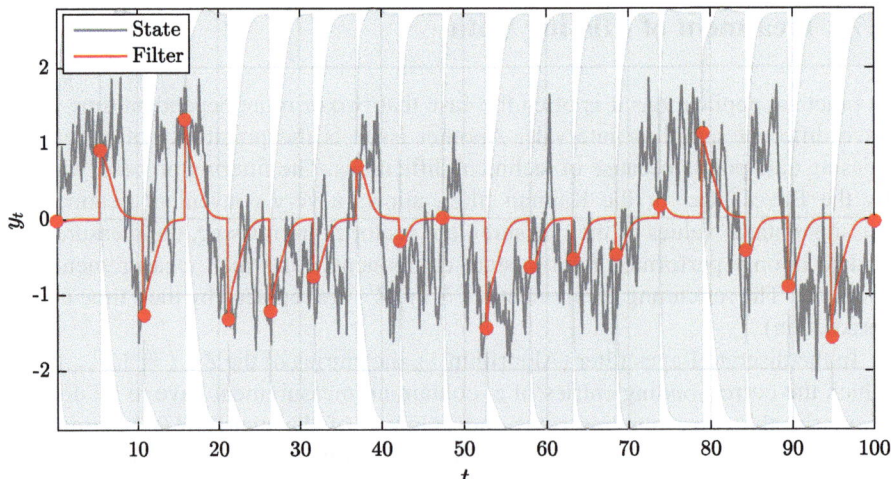

Fig. 2.5 Missing data and prediction

2.7 Treatment of Missing Data

gridpoints), 20 equidistant measurements and the initial a priori moments

$$\mu_{y,1|0} = \begin{bmatrix} 0 \\ 0 \end{bmatrix}, \Sigma_{yy,1|0} = \begin{bmatrix} 1 & 0 \\ 0 & 1 \end{bmatrix}.$$

Between measurements, the filter value is equal to the h-step prognosis $\mathbb{E}\left[y_{t-1+h\cdot\Delta t}|\mathbf{Z}^{t-1}\right]$ and, over time, approaches the unconditional mean value $\mathbb{E}[y_t] = 0$. The impact of missing data is also very clearly illustrated by the 99% HPD region, represented by the gray area. Between measurements, the area expands what is to be interpreted as increasing forecast uncertainty due to lack of measurements. Accordingly, the HPD region decreases abruptly when a new measurement arrives.

Chapter 3
Deterministic Numerical Integration

Often it is not possible to evaluate an integral analytically. This case may occur, for example, if the needed antiderivative is not known or the function $f(x)$ cannot be stated in a closed form at all. This chapter focuses on the description of methods for the approximation of integrals of the form

$$\mathcal{I}[fw] = \int_{I_1} \ldots \int_{I_d} f(x_1, \ldots, x_d) \, w(x_1, \ldots, x_d) \, dx_d \ldots dx_1, \tag{3.1}$$

with $I_1, \ldots, I_d \subseteq \mathbb{R}$.

Furthermore it is assumed that

$$w(x_1, \ldots, x_d) = w(x_1) \cdot \ldots \cdot w(x_d). \tag{3.2}$$

In order to shorten the notation if necessary, frequently the following abbreviation is going to be used:

$$\mathcal{I}[fw] = \int_D f(x) \, w(x) \, dx, \tag{3.3}$$

with $D \subseteq \mathbb{R}^d$.

The function $f(x)$ and the weight function $w(x)$ are supposed to be Riemann integrable on the domain of integration.

The methods of deterministic numerical integration used in this work are called quadrature, for the one-dimensional case and cubature, for the multidimensional case. The central concept here is the approximation of the function $f(x)$ by an interpolating polynomial. A fundamental property of the methods is that they are constructed in a way, so that the integral of the weighted interpolating polynomial can be evaluated *exactly* by calculating the sum

$$\sum_{l=1}^{n} f(\boldsymbol{\chi}_l)\alpha_l, \tag{3.4}$$

in which the points $\boldsymbol{\chi}_l$ are the called cubature abscissae and the α_l are the cubature weights. A particular integration method has a degree of exactness m, if it renders exact results for all weighted polynomials of degree up to m (cf. Gil et al. 2007, p. 124). To clarify the difference between cubature and quadrature in terms of notation, the abscissae of a quadrature rule in the following will be denoted by the non-bold symbol χ_l and the weights by ϑ_l. Decimal numbers generally will be given rounded to the second digit after the decimal point.

3.1 One-Dimensional Deterministic Numerical Integration

The principle of deterministic numerical integration is best illustrated by the case $d = 1$. A quadrature rule Q consists of a set of abscissae χ_l and weights ϑ_l, $l = 1, \ldots, n$. The approximation of the integral can now be formulated as

$$\int_a^b f(x)\,w(x)\,dx \approx \sum_{l=1}^{n} f(\chi_l)\,\vartheta_l =: Q^{m,n}[fw]. \tag{3.5}$$

The expression $Q^{m,n}[fw]$ means that the quadrature rule applied to $f(x)\,w(x)$ uses n abscissae and is exact for all polynomials up to degree m. Indeed, the quadrature rules used in this work possess the property that their quadrature value $Q^{m,n}[fw]$ convergences to the true value of the integral as the number of abscissae and weights are increased, so

$$\int_a^b f(x)\,w(x)\,dx = \lim_{n \Rightarrow \infty} Q^{m,n}[fw]. \tag{3.6}$$

3.1.1 Lagrange Interpolation

Before talking about how to numerically integrate the composition $f(x)\,w(x)$, it will prove useful to introduce the Lagrange interpolation functions first. They are defined as:

$$\varsigma_l(x) = \prod_{\substack{j=1 \\ j \neq l}}^{n} \frac{x - \chi_j}{\chi_l - \chi_j}, \quad l = 1, \ldots, n. \tag{3.7}$$

3.1 One-Dimensional Deterministic Numerical Integration

With their help the interpolant $\widehat{f}(x)$, which interpolates to the function $f(x)$ at the nodes

$$(\chi_1, f(\chi_1)), (\chi_2, f(\chi_2)), \ldots, (\chi_n, f(\chi_n)), \tag{3.8}$$

can be written as $\widehat{f}(x) = \sum_{l=1}^{n} f(\chi_l) \varsigma_l(x)$. After inspecting the Lagrange polynomials it becomes obvious that

$$\varsigma_l(\chi_k) = \begin{cases} 1, \text{ if } l = k \\ 0, \text{ if } l \neq k \end{cases}.$$

and that therefore the interpolant fulfills the requirement $\widehat{f}(\chi_l) = f(\chi_l)$ for $l = 1 \ldots n$. An important property of $\widehat{f}(x)$ is its uniqueness. The interpolating polynomial is of degree smaller than n and there is *no* other polynomial of degree smaller than n which passes through the given nodes (cf. Gil et al. 2007, p. 52). So if $f(x)$ itself is a polynomial and has degree less than n, then it follows that $f(x) = \widehat{f}(x)$ and therefore

$$\int_a^b f(x) w(x) \, dx = \int_a^b \widehat{f}(x) w(x) \, dx = \int_a^b \sum_{l=1}^{n} f(\chi_l) \varsigma_l(x) w(x) \, dx$$

$$= \sum_{l=1}^{n} f(\chi_l) \int_a^b \varsigma_l(x) w(x) \, dx. \tag{3.9}$$

This reveals that under the stated prerequisites the quadrature weights are

$$\vartheta_l = \int_a^b \varsigma_l(x) w(x) \, dx, \ l = 1 \ldots n. \tag{3.10}$$

Consequently, every weighted polynomial of degree smaller than n can be exactly evaluated by using the abscissae χ_l and the weights ϑ_l. Moreover it is important to note that in this approach, although it makes sense to choose the abscissae from the interval $[a, b]$, the positions of the n abscissae χ_l are totally arbitrary.

3.1.2 Moment Equations for the One-Dimensional Case

In dependence on the choice of the abscissae χ_l, quadrature rules exists which are capable of exactly integrating weighted polynomials of degree significantly higher than $n - 1$ by using n abscissae. Moreover, quadrature rules using n abscissae have been developed, which have special features but are only exact for polynomials of degree much smaller than $n-1$. This shows that various combinations of amounts of abscissae and degrees of exactness are possible.

The general problem of exactly integrating weighted one-dimensional polynomials of degree m by using n abscissae and weights can be vividly represented by the so-called moment equations. In order to construct a quadrature rule which is capable of integrating all weighted polynomials of degree up to m, the abscissae weights χ_l and associated weights ϑ_l must be determined. Polynomials of degree up to m are sums of the monomials $x^0, x^1, \ldots, x^m$ and therefore the abscissae and weights must fulfill the following m equations:

$$\sum_{l=1}^{n} \chi_l^i \vartheta_l = \int_a^b x^i w(x)\, dx, \; i = 0, \ldots, m. \qquad (3.11)$$

For a given set of abscissae, the linear equations from which the weights are derived (3.11) can be summarized by the following equation system:

$$\begin{bmatrix} 1 & 1 & \cdots & 1 \\ \chi_1 & \chi_2 & \cdots & \chi_n \\ \chi_1^2 & \chi_2^2 & \cdots & \chi_n^2 \\ \chi_1^3 & \chi_2^3 & \cdots & \chi_n^3 \\ \vdots & \vdots & \vdots & \vdots \\ \chi_1^m & \chi_2^m & \cdots & \chi_n^m \end{bmatrix} \cdot \begin{bmatrix} \vartheta_1 \\ \vartheta_2 \\ \vartheta_3 \\ \vartheta_4 \\ \vdots \\ \vartheta_n \end{bmatrix} = \begin{bmatrix} \int_a^b w(x)\, dx \\ \int_a^b x w(x)\, dx \\ \int_a^b x^2 w(x)\, dx \\ \int_a^b x^3 w(x)\, dx \\ \vdots \\ \int_a^b x^m w(x)\, dx \end{bmatrix} \qquad (3.12)$$

or in short notation

$$A\vartheta = b. \qquad (3.13)$$

According to the theorem of Rouché-Capelli (cf. Shafarevich and Remizov 2013, pp. 56–57), a system of linear equations has at least one solution, if the matrix A and the augmented Matrix $[A\;b]$ are of the same rank. Furthermore, if the system fulfills that condition and the number of unknowns is equal to the rank of the matrix A, then the system of equations has a unique solution. Otherwise, infinitely many solutions exist.

As can be assumed, various scenarios can occur, only some of which will be referred to here. If the number of rows, $m+1$, corresponds to the number of columns n and the matrix A has full rank, there always exists a unique solution in form of the vector ϑ, so that the above equality holds. The solution to the system of equations is then indeed equivalent to (3.10). If the system of equations is overdetermined ($m+1 > n$) and the columns are pairwise linearly independent, there might exist a unique solution or no solution. The latter is usually the case. In the third setting it is $m+1 < n$ which is referred to as underdetermination. If the rows are pairwise linearly independent it follows that infinitely many solutions exist for the system of equations. In each of these solutions at most m weights ϑ_l are different from zero.

3.1 One-Dimensional Deterministic Numerical Integration

A quadrature rule is called *interpolatory quadrature rule*, if its weights ϑ_l satisfy the above moment equations. An interpolatory quadrature rule can be interpreted as the exact integral of the polynomial which interpolates $f(x)$ at the abscissae χ_l (cf. Krommer and Ueberhuber 1998, p. 121).

If the integral

$$\mathcal{I}\left[x^j \cdot w(x)\right] = \int_a^b x^j w(x)\, dx \qquad (3.14)$$

vanishes whenever j is odd, $\mathcal{I}\left[x^j \cdot w(x)\right]$ is called centrally symmetric (cf. Möller 1976, p. 186). In this case the number of equations can be reduced. This will now be illustrated using the example of the standard normal distribution as weight function,

$$w(x) = \frac{1}{\sqrt{2\pi}} e^{-\frac{x^2}{2}}, \qquad (3.15)$$

with the domain of integration $\mathbb{R}$. Because $\int_{-\infty}^{\infty} x^j \frac{1}{\sqrt{2\pi}} e^{-\frac{x^2}{2}}\, dx = 0$ for $j = 2k+1$, $k = 0, 1, 2 \ldots$ the abscissae χ_l shall be defined the following way, in which n is set to an odd number:

$$\chi_1 = -\chi_n,\ \chi_2 = -\chi_{n-1},\ \chi_3 = -\chi_{n-2},\ \ldots,\ \text{and}\ \chi_{\frac{n-1}{2}+1} = 0 \qquad (3.16)$$

The weights ϑ_l then have the structure

$$\vartheta_1 = \vartheta_n,\ \vartheta_2 = \vartheta_{n-1},\ \vartheta_3 = \vartheta_{n-2},\ \ldots \qquad (3.17)$$

These restrictions ensure that

$$\left[\chi_1^j\ \chi_2^j\ \ldots\ \chi_n^j\right] \cdot \left[\vartheta_1\ \vartheta_2\ \ldots\ \vartheta_n\right]' = 0,\ \text{for } j = 2k+1,\ k = 0, 1, 2 \ldots. \qquad (3.18)$$

For $m = 4$ and $n = 5$, the system of equations (3.12) takes on the form

$$\begin{bmatrix} 1 & 1 & 1 & 1 & 1 \\ \chi_1 & \chi_2 & \chi_3 & \chi_4 & \chi_5 \\ \chi_1^2 & \chi_2^2 & \chi_3^2 & \chi_4^2 & \chi_5^2 \\ \chi_1^3 & \chi_2^3 & \chi_3^3 & \chi_4^3 & \chi_5^3 \\ \chi_1^4 & \chi_2^4 & \chi_3^4 & \chi_4^4 & \chi_5^4 \end{bmatrix} \cdot \begin{bmatrix} \vartheta_1 \\ \vartheta_2 \\ \vartheta_3 \\ \vartheta_4 \\ \vartheta_5 \end{bmatrix} = \begin{bmatrix} 1 \\ 0 \\ 1 \\ 0 \\ 3 \end{bmatrix}. \qquad (3.19)$$

After introducing the described restrictions, the equations change to

$$\begin{bmatrix} 1 & 1 & 1 & 1 & 1 \\ \chi_1 & \chi_2 & 0 & -\chi_2 & -\chi_1 \\ \chi_1^2 & \chi_2^2 & 0 & \chi_2^2 & \chi_1^2 \\ \chi_1^3 & \chi_2^3 & 0 & (-\chi_2)^3 & (-\chi_1)^3 \\ \chi_1^4 & \chi_2^4 & 0 & \chi_2^4 & \chi_1^4 \end{bmatrix} \cdot \begin{bmatrix} \vartheta_1 \\ \vartheta_2 \\ \vartheta_3 \\ \vartheta_2 \\ \vartheta_1 \end{bmatrix} = \begin{bmatrix} 1 \\ 0 \\ 1 \\ 0 \\ 3 \end{bmatrix}. \quad (3.20)$$

The equations in which the abscissae have odd exponents can be deleted, due to equation (3.18),

$$\begin{bmatrix} 1 & 1 & 1 & 1 & 1 \\ \chi_1^2 & \chi_2^2 & 0 & \chi_2^2 & \chi_1^2 \\ \chi_1^4 & \chi_2^4 & 0 & \chi_2^4 & \chi_1^4 \end{bmatrix} \cdot \begin{bmatrix} \vartheta_1 \\ \vartheta_2 \\ \vartheta_3 \\ \vartheta_2 \\ \vartheta_1 \end{bmatrix} = \begin{bmatrix} 1 \\ 1 \\ 3 \end{bmatrix}. \quad (3.21)$$

Finally, the system of equations is rewritten to

$$\begin{bmatrix} 2 & 2 & 1 \\ 2\chi_1^2 & 2\chi_2^2 & 0 \\ 2\chi_1^4 & 2\chi_2^4 & 0 \end{bmatrix} \cdot \begin{bmatrix} \vartheta_1 \\ \vartheta_2 \\ \vartheta_3 \end{bmatrix} = \begin{bmatrix} 1 \\ 1 \\ 3 \end{bmatrix}. \quad (3.22)$$

In addition to the Gaussian weight function, the weight function $w(x) = 1$ is used in this work. Here the exact same approach can be applied with the only difference that the results of the integrals have to be replaced. For $m = 4$, $w(x) = 1$ and the domain of integration $[-1, 1]$, the system of equations reads

$$\begin{bmatrix} 2 & 2 & 1 \\ 2\chi_1^2 & 2\chi_2^2 & 0 \\ 2\chi_1^4 & 2\chi_2^4 & 0 \end{bmatrix} \cdot \begin{bmatrix} \vartheta_1 \\ \vartheta_2 \\ \vartheta_3 \end{bmatrix} = \begin{bmatrix} 2 \\ 2/3 \\ 2/5 \end{bmatrix}. \quad (3.23)$$

Due to the restrictions and the central symmetry of the weight function, the produced quadrature rule is not only of degree $m = 4$, but also $m = 5$. Thus, using this approach, quadrature rules of odd degree m can be constructed by choosing $(m-1)/2$ abscissae and solving $(m-1)/2 + 1$ equations. If the abscissae are chosen so that the matrix A is of full rank, (3.23) always has a unique solution. A problem which must be taken into account is that those kinds of equation systems, because of the exponential multiplications, become unstable when m increases. Therefore, the results obtained should be carefully analysed. The resulting quadrature rules employ n abscissae and have a degree of exactness of $m = n$. In the next subsections it will be shown that by using special sets of abscissae also higher degrees of exactness can be achieved.

3.1.3 Gauss Quadrature

The core of Gauss quadrature is to choose the n abscissae and associated weights in a way that the constructed quadrature rule is not only exact for polynomials up to degree $n - 1$, but also for polynomials up to degree $2n - 1$ actually.

Presuming that $f(x)$ is a polynomial of degree $2n - 1$, after Euclidean division of $f(x)$ by a polynomial $p(x) \in \mathbb{P}_n$[1] one can write $f(x) w(x)$ as (cf. Vialar 2015, p. 110)

$$\underbrace{f(x)}_{\in \mathbb{P}_{2n-1}} w(x) = \left(\underbrace{q(x)}_{\in \mathbb{P}_{n-1}} \cdot \underbrace{p(x)}_{\in \mathbb{P}_n} + \underbrace{r(x)}_{\in \mathbb{P}_{n-1}} \right) w(x) \quad (3.24)$$

$$= q(x) p(x) w(x) + r(x) w(x).$$

The polynomial $p(x)$ is now chosen so that it meets the following condition:

$$\int_a^b q(x) p(x) w(x) \, dx = 0, \text{ for all polynomials } q(x) \in \mathbb{P}_{n-1}. \quad (3.25)$$

In analogy to the scalar product of vectors the scalar product of two polynomials $q(x)$ and $p(x)$ with respect to a weight function $w(x)$ on the domain $[a, b]$ may be defined as

$$\langle q(x), p(x) \rangle \equiv \int_a^b q(x) p(x) w(x) \, dx. \quad (3.26)$$

Accordingly, equation (3.25) indicates that the polynomial $p(x)$ is orthogonal to all polynomials $q(x)$ of degree less than n with respect to the weight function $w(x)$. For the integration of $f(x) w(x)$ it now follows that

$$\int_a^b f(x) w(x) \, dx = \underbrace{\int_a^b q(x) p(x) w(x) \, dx}_{=0, \text{ due to orthogonality}} + \int_a^b r(x) w(x) \, dx \quad (3.27)$$

$$= \int_a^b r(x) w(x) \, dx.$$

So the composition $q(x) p(x) w(x)$ vanishes automatically and only $r(x) w(x)$ remains. As mentioned previously, the rest polynomial $r(x)$ has degree lower than n and hence $r(x) w(x)$ can be integrated by means of interpolatory quadrature (3.9).

---

[1] $\mathbb{P}_n$ is the $n + 1$-dimensional polynomial vector space and is spanned by the monomial basis $\{1, x, x^2, \ldots, x^n\}$.

The Gauss quadrature takes advantage of this relation. Assume the polynomial $p(x)$ to be known. Then it can be resolved into linear factors:

$$p(x) = (\chi_1 - x) \cdot (\chi_2 - x) \cdot \ldots \cdot (\chi_n - x). \tag{3.28}$$

The parameters χ_l represent the n roots of the polynomial. Choosing the zeros of the polynomial $p(x)$ as abscissae for the Gauss quadrature yields the desired result. It turns out that

$$\begin{aligned}
\sum_{l=1}^{n} f(\chi_l) \vartheta_l &= \sum_{l=1}^{n} q(\chi_l) \cdot p(\chi_l) \vartheta_l + \sum_{l=1}^{n} r(\chi_l) \vartheta_l \\
&= \underbrace{\sum_{l=1}^{n} q(\chi_l) \cdot (\chi_1 - \chi_l) \cdot \ldots \cdot (\chi_n - \chi_l) \vartheta_l}_{=0} + \sum_{l=1}^{n} r(\chi_l) \vartheta_l \\
&= \sum_{l=1}^{n} r(\chi_l) \vartheta_l, \text{ with } \vartheta_l = \int_a^b \varsigma_l(x) w(x) dx \\
&= \int_a^b r(x) w(x) dx = \int_a^b f(x) w(x) dx.
\end{aligned} \tag{3.29}$$

A higher degree than $2n + 1$ cannot be reached, because otherwise the orthogonal polynomial $p(x)$ would also have to be orthogonal to itself, what is impossible.

The task is now to find the orthogonal polynomial $p(x)$, in case it exists. A system of orthogonal polynomials can be constructed from the monomial basis $\{1, x, x^2, \ldots\}$ by using the Gram–Schmidt orthogonalization algorithm (Schmidt 1908). This procedure leads to a set of monic polynomials what means that the leading coefficient of each polynomial is one. The nth orthogonal polynomial has exactly n distinct roots on the interval $[a, b]$ (cf. Gubner 2009, pp. 2–3). Furthermore, from the orthogonality property it follows that

$$\langle p_l(x), p_k(x) \rangle = \begin{cases} 0, & \text{if } l \neq k \\ \neq 0, & \text{if } l = k \end{cases}. \tag{3.30}$$

The construction proceeds as follows:

$$p_0(x) := 1$$
$$p_n(x) := x^n - \sum_{l=0}^{n-1} \frac{\langle x^n, p_l(x) \rangle}{\langle p_l(x), p_l(x) \rangle} p_l(x). \tag{3.31}$$

The predefined p_0 together with the resulting polynomials $p_1(x), p_2(x), \ldots, p_n(x)$ also form basis for $\mathbb{P}_n$. In addition the just mentioned polynomials even form a

3.1 One-Dimensional Deterministic Numerical Integration

orthogonal basis for $\mathbb{P}_n$ with respect to the weight function $w(x)$. Summarized, for the numerical calculation of the integral $\int_a^b f(x) w(x) \, dx$, where $f(x)$ is a polynomial of degree at most $2n - 1$, one needs

1. to find the roots of the nth orthogonal polynomial and
2. to calculate the weights ϑ_l.

It has to be noted that not with respect to every weight function $w(x)$ a Gauss quadrature exists. Nevertheless it can be proven that the existence is guaranteed, if $w(x) > 0$. Then, the weights ϑ_l lie inside the interval $[a, b]$ and are positive (cf. Gautschi 1968; Engels 1980, pp. 202–206).

There are several methods available for the calculation of the zeros, respectively, abscissae χ_l and weights ϑ_l of a Gauss quadrature rule. A straightforward procedure is to construct the needed polynomials using the Gram–Schmidt orthogonalization algorithm and then to determine the zeros analytically. Afterwards the weights can be obtained by evaluating the integrals (3.10). A very elegant way is provided in form of the Golub–Welsch algorithm (Golub and Welsch 1969). It transforms the task to an eigenvalue problem and is particularly efficient if the considered weight function stems from the class of the so-called classical weight functions. The weight functions which lead to the Legendre and Hermite polynomials, which are to be discussed in the following sections, also belong to the class of classical weight functions. An explanation of the algorithm is given in appendix (C).

Gauss–Legendre Quadrature Gauss–Legendre quadrature is suitable in the case, where an integral of the form $\int_a^b f(x) \, dx$ has to be calculated. Accordingly, $w(x) = 1$. The integration interval $[a, b]$ is set to $[-1, 1]$ by convention and the abscissae and weights are tabulated for this specific domain. This is an unproblematic restriction, because the integration variable of any integral can always be substituted in a way that the integration is carried out over a desired region (see, for instance, Sect. 3.2.8).

The orthogonal polynomials needed to carry out the numerical integration are called Legendre polynomials $L_l(x)$. In order to illustrate the construction of the Legendre polynomials, the calculations of $L_1(x)$, $L_2(x)$ and $L_3(x)$ are made concrete in (3.32), (3.33) and (3.34):

$$L_1(x) := x - \frac{\int_{-1}^{1} x \, dx}{\int_{-1}^{1} 1 \, dx} = x \tag{3.32}$$

$$L_2(x) := x^2 - \frac{\int_{-1}^{1} x^2 \, dx}{\int_{-1}^{1} 1 \, dx} - \frac{\int_{-1}^{1} x^2 \cdot x \, dx}{\int_{-1}^{1} x^2 \, dx} x = x^2 - \frac{1}{3} \tag{3.33}$$

$$L_3(x) := x^3 - \frac{\int_{-1}^1 x^3\, dx}{\int_{-1}^1 1\, dx} - \frac{\int_{-1}^1 x^3 \cdot x\, dx}{\int_{-1}^1 x^2\, dx} x$$

$$- \frac{\int_{-1}^1 x^3 \cdot \left(x^2 - \frac{1}{3}\right) dx}{\int_{-1}^1 \left(x^2 - \frac{1}{3}\right) \cdot \left(x^2 - \frac{1}{3}\right) dx} \cdot \left(x^2 - \frac{1}{3}\right) = x^3 - \frac{3}{5}x \quad (3.34)$$

Gauss–Hermite Quadrature In the case where $w(x) = \frac{1}{\sqrt{2\pi}} e^{-\frac{x^2}{2}}$ and $a = -\infty$ and $b = \infty$, Gauss–Hermite quadrature can be applied. The abscissae in this case are the roots of the so-called (probabilists') Hermite polynomials He_l (cf. Abramowitz and Stegun 1972, pp. 775). As the Legendre polynomials, these orthogonal polynomials can be constructed from the monomial basis, however, the weight function $w(x)$ must be taken into account.

Again, as illustrative example, the calculations of $He_1(x)$, $He_2(x)$ and $He_3(x)$ according to the Gram–Schmidt orthogonalization algorithm (3.31) shall be carried out in detail:

$$H_1(x) := x - \frac{\int_{-\infty}^\infty x \cdot e^{-\frac{x^2}{2}}\, dx}{\int_{-\infty}^\infty e^{-\frac{x^2}{2}}\, dx} = x \quad (3.35)$$

$$H_2(x) := x^2 - \frac{\int_{-\infty}^\infty x^2 \cdot e^{-\frac{x^2}{2}}\, dx}{\int_{-\infty}^\infty e^{-\frac{x^2}{2}}\, dx} - \frac{\int_{-\infty}^\infty x^2 \cdot x \cdot e^{-\frac{x^2}{2}}\, dx}{\int_{-\infty}^\infty x^2 \cdot e^{-\frac{x^2}{2}}\, dx} x = x^2 - 1$$

$$(3.36)$$

$$H_3(x) := x^3 - \frac{\int_{-\infty}^\infty x^3 \cdot e^{-\frac{x^2}{2}}\, dx}{\int_{-\infty}^\infty e^{-\frac{x^2}{2}}\, dx} - \frac{\int_{-\infty}^\infty x^3 \cdot x \cdot e^{-\frac{x^2}{2}}\, dx}{\int_{-\infty}^\infty x^2 \cdot e^{-\frac{x^2}{2}}\, dx} x$$

$$- \frac{\int_{-\infty}^\infty x^3 \cdot \left(x^2 - \frac{1}{2}\right) \cdot e^{-\frac{x^2}{2}}\, dx}{\int_{-\infty}^\infty \left(x^2 - \frac{1}{2}\right) \cdot \left(x^2 - \frac{1}{2}\right) \cdot e^{-\frac{x^2}{2}}\, dx} \cdot \left(x^2 - \frac{1}{2}\right) \quad (3.37)$$

$$= x^3 - 3x$$

Gauss–Kronrod Quadrature A big disadvantage of the Gauss quadrature is the non-nestedness of different sets of abscissae. Strictly speaking, two sets of abscissae belonging to different degrees of exactness share the midpoint *at most*. This drawback has the effect that the error of the used Gauss quadrature rule of specific polynomial exactness m cannot be quantified efficiently. As an example, the error estimate for the Gauss quadrature rule $Q^{m,n}[fw]$ shall be considered. The rule uses n abscissae and is of exactness $m = 2n - 1$. A natural error estimate would be constructed by increasing the number of points to $n + 1$, so that $m = 2n + 1$ and

3.1 One-Dimensional Deterministic Numerical Integration

taking the absolute value of the difference of the two results. So the error measure would be

$$Q^{2n-1,n} [fw]_{\text{Err}} = \left| Q^{2n+1,n+1} [fw] - Q^{2n-1,n} [fw] \right|. \tag{3.38}$$

This measure of error is not very efficient. The procedure needs the evaluation of $2n + 1$ function values, but the reference quadrature $Q^{2n+1,n+1} [fw]$ is only of exactness $m = 2n + 1$ and is therefore not much more exact than $Q^{2n-1,n} [fw]$. Especially if $f(x)$ is not a smooth function, according to Monegato (2001), the reference Gauss quadrature rule should have twice as many abscissae as the quadrature rule whose error is to be estimated. The Gauss–Kronrod quadrature (Kronrod 1965) solves this problem to a certain extent and will now be described.

The idea is to take the n abscissae of the Gauss quadrature rule and to add $n + 1$ new abscissae. The result is a quadrature rule with n preassigned and $n + 1$ new abscissae:

$$Q^{3n+1,2n+1} [fw] = \sum_{l=1}^{n} \vartheta_l f(\chi_l) + \sum_{k=1}^{n+1} \beta_k f(\eta_k). \tag{3.39}$$

The parameters β_k and η_k represent the Gauss–Kronrod weights and abscissae. The weights ϑ_l cannot be reused in this approach, so all $2n + 1$ weights have to be calculated from scratch. This approach results in a quadrature rule of exactness $m = 2n + 1 + c$, in which the function to be integrated only has to be evaluated at $n + 1$ new abscissae. In order to determine the degree of exactness, the highest possible value for c has to be investigated.

Analogous to (3.24), if $f(x)$ is a weighted polynomial of degree $2n + 1 + c$, it can be written as

$$\underbrace{f(x)}_{\in \mathbb{P}_{2n+1+c}} w(x) = \left(\underbrace{q(x)}_{\in \mathbb{P}_c} \cdot \underbrace{p(x)}_{\in \mathbb{P}_{2n+1}} + \underbrace{r(x)}_{\in \mathbb{P}_{2n}} \right) w(x) \tag{3.40}$$

$$= q(x) p(x) w(x) + r(x) w(x).$$

In this case, $p(x)$ is the polynomial of degree $2n+1$ which has the n abscissae of the used Gauss quadrature rule and the desired $n+1$ Gauss–Kronrod abscissae as roots. $r(x)$ is the usual rest polynomial which has degree smaller than $2n + 1$. Again, the polynomial $p(x)$ must be chosen, so that

$$\int_a^b q(x) p(x) w(x) \, dx = 0, \text{ for all polynomials } q(x) \in \mathbb{P}_c. \tag{3.41}$$

Because n roots of $p(x)$ are preassigned, one can write the decomposition

$$p(x) = \pi_{n+1}(x) \prod_{l=1}^{n}(x - \chi_l) = \pi_{n+1}(x) z_n(x). \tag{3.42}$$

The last equality holds, because every polynomial of degree n can be uniquely written as the product of n linear factors where the χ_l are the roots of the polynomial. The polynomial $z_n(x)$ is orthogonal to all polynomials $\in \mathbb{P}_{n-1}$. In the case of $w(x) = 1$, $z_n(x)$ is the Legendre polynomial $L_n(x)$. Inserting the decomposition into (3.41) yields

$$\int_a^b q(x) \pi_{n+1}(x) z_n(x) w(x) \, dx = 0$$

$$\int_a^b q(x) \pi_{n+1}(x) \underbrace{z_n(x) w(x)}_{\tilde{w}(x)} dx = 0, \text{ for } q(x) \in \mathbb{P}_c. \tag{3.43}$$

As can be seen, the product of the known component of the decomposition (3.42) and the weight function $w(x)$ are to be interpreted as new composite weight function $\tilde{w}(x)$. Thus, the polynomial $\pi_{n+1}(x)$, called Stieltjes polynomial (cf. Gautschi and Notaris 1996), must be orthogonal to all polynomials $\in \mathbb{P}_c$ with respect to the composite weight function $\tilde{w}(x)$ and the roots of this polynomial serve as the Gauss–Kronrod abscissae β_k. Because the Stieltjes polynomial is $\in \mathbb{P}_{n+1}$ it can *at most* be orthogonal to all polynomials $\in \mathbb{P}_n$ with respect to the composite weight function. Therefore, it is $c = n$ and the degree of exactness of the Gauss–Kronrod quadrature generally is $m = 2n + 1 + n = 3n + 1$.

It can be shown (cf. Rabinowitz 1980) that the degree of exactness varies for some special cases of weight functions, e.g. for $w(x) = 1$ the exact degree of polynomial exactness is $3n + 1$ for even n and $3n + 2$ for odd n. So with respect to the Gauss–Kronrod quadrature, the value $m = 3n + 1$ is meant to be the lower bound of polynomial exactness.

Some authors define the Gauss–Kronrod quadrature only to be existent if all abscissae are real and contained in the integration interval (cf. Ehrich 1995, p. 290). This definition will be adopted in the following. Although, from a practical point of view, the roots of $\pi_{n+1}(x)$ can be complex or lay outside the domain of integration, but these scenarios can easily lead to useless quadrature formulas (cf. Elhay and Kautsky 1992, p. 82). These cases will therefore not be considered further in this work. In the case of Gauss–Hermite quadrature, Gauss–Kronrod rules are only existent for $n = 1, 2, 4$ (cf. Gil et al. 2007, p. 300). For weight functions of the form $w_\lambda(x) = \left(1 - x^2\right)^{\lambda - \frac{1}{2}}$, $\lambda \in [0, 2]$ and $x \in [-1, 1]$ it has been proven by

3.1 One-Dimensional Deterministic Numerical Integration

Szegö (1935) that for all $n \in \mathbb{N}$ the zeros of $\pi_{n+1}(x)$ are real and moreover

- all lie inside the interval $[1, 1]$,
- are distinct,
- and interlace with the n zeros of $z_n(x)$.

For $\lambda \in [0, 1]$ Monegato (1978) has proven that the resulting weights are positive for all $n \in \mathbb{N}$. The positivity of the weights is not a mandatory requirement, but nevertheless desirable. Accordingly, the Gauss–Kronrod rule exists in particular for the Legendre weight function $\left(\lambda = \frac{1}{2}\right)$ with all-positive weights. The calculation of the $3n + 1$ values ϑ_l, β_k and η_k which are needed for the Gauss–Kronrod rule can be calculated in various ways. One convenient way in the case of positive weights has been proposed by Laurie (1997). He shows that the calculation of the abscissae and weights can be transformed to an eigenvalue problem to which the Golub–Welsch algorithm (Appendix C) is applicable.

Gauss–Patterson Quadrature The quadrature rule of Patterson (1968) is based on the idea of interleaving different Gauss–Kronrod rules. This means that, in the first step, starting from a Gauss rule with n abscissae, $n + 1$ abscissae are added to construct the Gauss–Kronrod rule with $n_P = n + (n + 1)$ abscissae. Now, another $n_p + 1$ abscissae are added and the Gauss–Kronrod method is applied again. The procedure is continued and therefore in each step of the iteration the amount of abscissae is doubled plus one. By doing so, nested sets of quadrature abscissae can be calculated:

$$Q^{3 \cdot 2^{i-1}(n+1) - 2, 2^i(n+1) - 1} = \sum_{l=1}^{n} \vartheta_{il} f(\chi_l) + \sum_{p=1}^{i} \sum_{l=1}^{2^{p-1}(n+1)} \beta_{ilp} f(\eta_{lp}). \tag{3.44}$$

An example will shed light on the complex formula for the Gauss–Patterson rule. The variable n represents the *basic* number of Gauss abscissae which are chosen as the basis to start the procedure. The variable n_P stands for the *current* number of abscissae and increases in each step of the calculation.

As an example, now the first two Gauss–Patterson quadrature rules for the Gaussian weight function, starting with one abscissae ($n = 1$), will be calculated. The polynomial $z_1(x)$ in this case is the first Hermite polynomial,

$$H_1(x) = x, \tag{3.45}$$

with the root 0. The procedure starts by constructing the first Gauss–Patterson rule, which is the Gauss–Kronrod rule. The polynomial $p(x) \in \mathbb{P}_3$ which satisfies

$$\int_{-\infty}^{\infty} q(x) p(x) w(x) = 0 \tag{3.46}$$

must have the root 0 and additionally two other roots, which are not yet known. Therefore, the polynomial can be decomposed into

$$p(x) = \pi_2(x)(x-0) = \pi_2(x)x = \pi_2(x)z_1(x). \qquad (3.47)$$

The integral equation then reads

$$\int_{-\infty}^{\infty} q(x)\pi_2(x)z_1(x)w(x) = 0. \qquad (3.48)$$

Gathering all fixed terms yields the new composite weight function

$$\tilde{w}(x) = z_1(x)w(x). \qquad (3.49)$$

Therefore, the task is to find the Stieltjes polynomial $\pi_2(x) \in \mathbb{P}_2$, so that

$$\int_{-\infty}^{\infty} q(x)\pi_2(x)\tilde{w}(x)\,dx = 0. \qquad (3.50)$$

for all polynomials $q(x) \in \mathbb{P}_1$. Applying the Gram–Schmidt orthogonalization algorithm (3.31) leads to

$$\pi_2(x) = x^2 - 1 \qquad (3.51)$$

with the roots -1.73 and 1.73. The resulting set includes $n_P = 3$ abscissae,

$$S^3 = \{-1.73, 0, 1.73\}, \qquad (3.52)$$

and coincides with the set of the first three Gauss–Hermite abscissae. The quadrature rule is exact for polynomials of degree $m \leq 3n + 1 = 4$ as already explained in Sect. 3.1.3. Indeed, this set is an exception to the general rule that $m = 3n + 1$ due to the fact that the first three Gauss–Hermite abscissae yield a quadrature rule of exactness $m = 5$.

For the construction of the second Gauss–Patterson extension, four new abscissae must be added to the set S^3 and as usual one must find the polynomial $p(x) \in \mathbb{P}_7$ which satisfies

$$\int_{-\infty}^{\infty} q(x)p(x)w(x) = 0. \qquad (3.53)$$

This time, three roots are fixed, namely $-1.73, 0$ and 1.73. The four remaining roots are not yet known. It follows the decomposition

$$p(x) = \pi_4(x)(x-0)(x+1.73)(x-1.73) = \pi_4(x)z_1(x)\pi_2(x). \qquad (3.54)$$

3.1 One-Dimensional Deterministic Numerical Integration

Separating all fixed components and defying the new compound weight function,

$$\tilde{w}(x) = z_1(x) \pi_2(x) w(x), \tag{3.55}$$

yields the requirement

$$\int_{-\infty}^{\infty} q(x) \pi_4(x) \tilde{w}(x) = 0. \tag{3.56}$$

Because the Stieltjes polynomial $\pi_4(x)$ is of degree $m = 4$ it can be at most be orthogonal to polynomials $q(x) \in \mathbb{P}_3$ with respect to $\tilde{w}(x)$. So $p(x) \in \mathbb{P}_7$, $q(x) \in \mathbb{P}_3$ and furthermore the rest polynomial $r(x)$ is always of smaller degree than the denominator polynomial $p(x)$ (cf. Vialar 2015, p. 110). Thus, the weighted polynomial $f(x)$ can therefore be rewritten to

$$\underbrace{f(x)}_{\in \mathbb{P}_{10}} w(x) = \Bigg(\underbrace{q(x)}_{\in \mathbb{P}_3} \cdot \underbrace{p(x)}_{\in \mathbb{P}_7} + \underbrace{r(x)}_{\in \mathbb{P}_6} \Bigg) w(x) \tag{3.57}$$

what indicates that the rule is exact for polynomials $\in \mathbb{P}_{10}$. Using the Gram–Schmidt orthogonalization algorithm (3.31), the following solution arises:

$$\pi_4(x) = x^4 - 10x^2 - 5. \tag{3.58}$$

The roots, $-3.24, -0.69i, 0.69i, 3.24$, are not entirely real and so this Gauss–Patterson rule is not usable for practical applications. Consequently, no further rules can be constructed.

Expressing the procedure in terms of one formula, the Stieltjes polynomial $\pi_{2^{i-1}(n+1)}(x)$ of the ith Gauss–Patterson rule must posses the following property:

$$\int_a^b \underbrace{z_n(x) w(x) \prod_{l=1}^{i-1} \pi_{2^{l-1}(n+1)}(x)}_{\tilde{w}(x)} \pi_{2^{i-1}(n+1)}(x) q(x) \, dx = 0, \tag{3.59}$$

for $q(x) \in \mathbb{P}_{2^{i-1}(n+1)-1}$.

The product arises due to the nesting of successive Gauss–Kronrod rules. For what cases the Patterson exist, depends on the choice of the parameter n and the weight function $w(x)$. Although some proofs for special scenarios have been published, many of the existing Gauss–Patterson rules have simply been found by computation. As already mentioned, in the case of Gauss–Hermite quadrature, Gauss–Kronrod rules can only be calculated for $n = 1, 2, 4$ (cf. Gil et al. 2007, p. 300). Consequently, Gauss–Patterson rules can only be constructed on the basis of these sets of abscissae. Unfortunately, these rules are complex even for small parameters n_p. Summarized, Gauss–Patterson quadrature in combination with the Hermite weight function is not useful for practical applications. In contrast, the usage in combination with the constant weight function $w(x) = 1$ is unproblematic. As an example, for $n = 1$ Patterson rules up to 511 abscissae have been successfully determined (cf. Mehrotra and Papp 2012, p. 3).

A Patterson-like method to construct nested sets in the case of the Hermite weight function has been proposed by Genz and Keister (1996). Their approach is based on finding that nested sets can be generated by constant adjustment of the variable n_P. In each step n_P has to be chosen in a way, so that no complex abscissae arise. One example is the following set. Starting with $n = 1$, Genz and Keister add the usual two Kronrod abscissae to the set. Until then, the procedure corresponds exactly to the Gauss–Kronrod, respectively, Gauss–Patterson quadrature. Now, instead of adding another 4 abscissae as required by the Patterson quadrature, they go on by adding 6 abscissae, which leads to non-complex roots and a degree of polynomial exactness of 15. In the next steps 10 and after that 16 abscissae are added, and so forth. So the number of additional abscissae is selected depending on whether the roots of the Stieltjes polynomial are complex or not.

3.1.4 Clenshaw–Curtis Quadrature

The Clenshaw–Curtis quadrature (Clenshaw and Curtis 1960) is based on the idea of approximating the function $f(x)$ by a Chebyshev series expansion. The employed polynomials are the Chebyshev polynomials of the first kind $T_i(x)$ which are defined as (cf. Mason and Handscomb 2003, p. 2)

$$T_i(x) = \cos[i \cdot \mathrm{acos}(x)], \quad x \in [-1, 1], \quad i = 0, 1, 2, \ldots \tag{3.60}$$

or

$$T_i[\cos(\varphi)] = \cos(i\varphi), \quad i = 0, 1, 2, \ldots \tag{3.61}$$

The polynomials $T_0(x), T_1(x), \ldots, T_n(x)$ are linear independent and form a basis of the polynomial vector space $\mathbb{P}_n$. Furthermore, the $T_i(x)$ are orthogonal to each other with respect to the weight function $w(x) = \frac{1}{\sqrt{1-x^2}}$. So for these polynomials, the scalar product may be defined as

$$\langle T_j(x), T_k(x) \rangle \equiv \int_{-1}^{1} T_j(x) T_k(x) \frac{1}{\sqrt{1-x^2}} dx. \tag{3.62}$$

3.1 One-Dimensional Deterministic Numerical Integration

In addition to the orthogonality property, two more properties can be stated (cf. Mason and Handscomb 2003, pp. 70–71). Summarized, the properties read

$$\langle T_j(x), T_k(x) \rangle = \begin{cases} 0, & \text{if } j \neq k \\ \pi, & \text{if } j = k = 0 \\ \frac{\pi}{2}, & \text{if } j = k \neq 0 \end{cases}. \tag{3.63}$$

The $n+1$ *distinct extrema* of $T_n(x)$, which will serve as quadrature abscissae, play an important role in the following explanations regarding the Clenshaw–Curtis quadrature. The $n+1$ abscissae χ_l are calculated as

$$\chi_l = \cos\left(\frac{l\pi}{n}\right), \quad l = 0, \ldots, n \tag{3.64}$$

and so, by (3.61),

$$T_j(\chi_l) = \cos\left(\frac{jl\pi}{n}\right). \tag{3.65}$$

Obviously, the abscissae χ_l are the extrema of T_n, because $T_n(\chi_l) = \cos(l\pi)$. With respect to this particular set of abscissae, $\{\chi_l\}_{l=0}^n$ and $j, k \leq n$, a special discrete orthogonality property can be derived by applying (3.65) (cf. Mason and Handscomb 2003, p. 84)[2]:

$$\sum_{l=0}^{n}{}'' T_j(\chi_l) T_k(\chi_l) = \begin{cases} 0, & \text{if } j \neq k \\ n, & \text{if } j = k = 0 \text{ or } j = k = n \\ \frac{n}{2}, & \text{if } 0 < j = k < n \end{cases}. \tag{3.66}$$

The double dash in $\sum''$ indicates that the first and the last summand are to be halved.

As usual, for the following derivation the function $f(x)$ shall be supposed to be a polynomial of degree n. Then it can be constructed by the Chebyshev polynomials up to degree n and be written as

$$f(x) = \sum_{j=0}^{n}{}'' c_j T_j(x)$$

$$\Rightarrow f(\chi_l) = \sum_{j=0}^{n}{}'' c_j T_j(\chi_l). \tag{3.67}$$

---
[2] After a lengthy derivation.

Multiplying both sides of the equation with $T_k(\chi_l)$ and summing this expression at all χ_l, yields

$$\sum_{l=0}^{n\;\prime\prime} f(\chi_l) T_k(\chi_l) = \sum_{j=0}^{n\;\prime\prime} c_j \left(\sum_{l=0}^{n\;\prime\prime} T_j(\chi_l) T_k(\chi_l) \right). \quad (3.68)$$

According to (3.66), the bracket on the right-hand side is only unequal to zero, if $k = j$. Thus, one finds

$$c_j = \frac{2}{n} \sum_{l=0}^{n\;\prime\prime} f(\chi_l) T_j(\chi_l) = \frac{2}{n} \sum_{l=0}^{n\;\prime\prime} f(\chi_l) \cos\left(\frac{jl\pi}{n}\right). \quad (3.69)$$

The integral of $f(x)$ is

$$\int_{-1}^{1} f(x) = \sum_{j=0}^{n\;\prime\prime} c_j \int_{-1}^{1} T_j(x)\, dx \quad (3.70)$$

$$= \sum_{j=0}^{n\;\prime\prime} \frac{2}{n} \sum_{l=0}^{n\;\prime\prime} f(\chi_l) \cos\left(\frac{jl\pi}{n}\right) \int_{-1}^{1} T_j(x)\, dx \quad (3.71)$$

$$= \sum_{l=0}^{n\;\prime\prime} f(\chi_l) \cdot \frac{2}{n} \sum_{j=0}^{n\;\prime\prime} \cos\left(\frac{jl\pi}{n}\right) \int_{-1}^{1} T_j(x)\, dx. \quad (3.72)$$

Using that

$$\int_{-1}^{1} T_j(x)\, dx = \begin{cases} 2/(1 - j^2), & \text{if } j \text{ is even} \\ 0, & \text{if } j \text{ is odd} \end{cases}, \quad (3.73)$$

one can finally write $\int_{-1}^{1} f(x)$ as $\sum_{l=0}^{n} f(\chi_l) \vartheta_l$. The weights ϑ_l are to be calculated according to the formula (cf. Waldvogel 2006, p. 197)

$$\vartheta_l = \frac{g_l}{n} \left[1 - \sum_{j=1}^{\lfloor n/2 \rfloor} \frac{b_j}{4j^2 - 1} \cos\left(\frac{2jl\pi}{n}\right) \right]^3 \quad (3.74)$$

with

$$g_l = \begin{cases} 1, & \text{if } l = 0 \text{ or } l = n \\ 2, & \text{if } 0 < l < n \end{cases}, \quad b_j = \begin{cases} 1, & \text{if } j = \frac{n}{2} \\ 2, & \text{if } j < \frac{n}{2} \end{cases}. \quad (3.75)$$

---

[3] The expression $\lfloor x \rfloor$ means that x is rounded to the largest integer not greater than x.

The abscissae χ_l and weights ϑ_l can be efficiently computed by using the discrete Fourier transform (DFT) (cf. Waldvogel 2006). When using the Clenshaw–Curtis quadrature one needs n abscissae and weights in order to exactly integrate polynomials of degree up to n for the case that n is odd. So in terms of polynomial exactness the procedure seems to be inferior to the Gauss quadrature. But actually it has been shown by Trefethen (2008) that for many practical applications in which the function $f(x)$ will mostly be drastically different from a polynomial, the performance of the Clenshaw–Curtis quadrature is comparable to the performance of the Gauss quadrature with the same number of used abscissae. This is due to the fact that the Chebyshev series expansion in many cases converges rapidly fast to the function $f(x)$ (cf. Olver, Frank W. J. 2010, p. 97). Another key advantage is of great use: The Clenshaw–Curtis quadrature can be used to form a family of fully nested quadrature rules. Two Clenshaw–Curtis quadrature rules of different exactness which use n_i and n_{i+1} abscissae are always nested, if one defines $n_1 = 1$ and $n_i = 2^{(i-1)} + 1$ for $i > 1$. Therefore, in this case, it is $S^{n_i} \subset S^{n_{i+1}}$. As a consequence, to reach a higher degree of exactness, the already computed function values to the set S^{n_i} can be reused and only the function values for the difference set $\{S^{n_{i+1}} \setminus S^{n_i}\}$ have additionally to be evaluated. This is not the case for the Gauss quadrature rules, in which two sets S^{n_i} and $S^{n_{i+1}}$ at most have the midpoint in common.

3.2 Multidimensional Deterministic Numerical Integration

If the function which has to be integrated is multidimensional, $d \geq 2$, one speaks of cubature instead of quadrature. Like in the one-dimensional case, the exactness of the employed cubature rule is measured in terms of its polynomial exactness. A general problem connected to the field of multidimensional numerical integration is the circumstance that the number of abscissae χ_l needed to approximate integrals up to a level of exactness m often tends to increase strongly as the dimension d rises. This makes the evaluation of the cubature rule extremely time consuming. Therefore a main focus of research in recent decades has been the development of formulas that use a small number of abscissae.

To mention only a few, regarding various kinds of weight functions and integration regions important results have been found by Stroud and Secrest (1963), Stenger (1971), Haegemans and Piessens (1976), Beckers and Haegemans (1991), Lyness and Cools (1994), Genz and Keister (1996), Cools et al. (1998), Petras (2003), Victoir (2004), Kuperberg (2006) and Hinrichs and Novak (2007). The book of Stroud (1971) is, despite its age, a valuable compendium, containing cubature rules which are still relevant today. The work of Stroud is continued by Cools (Cools and Rabinowitz 1993; Cools 1999) who has created a website which gives an overview of most of the known cubature rules regarding various regions of integration and weights functions of huge practical relevance (Cools 2003).[4]

Regarding the regions and weight functions which are of interest in this work, various efficient cubature rules are already known. Some of them use strictly positive weights but are only applicable to a fixed dimension d and a fixed degree of exactness m. Others are valid for a specified m and arbitrary dimension d, with weights being negative from a certain d on. A third class of cubature rules is usable for arbitrary d and m and uses negative weights for the majority of combinations of d and m. Furthermore, the influence of the negative weights increases with rising d and m. Whereas in the field of one-dimensional integration numerous efficient rules with all-positive weights and arbitrary degree of exactness are available for a wide range of integration problems, for the kinds of integrals which are of interest in this work, no general method for the construction of cubature rules for arbitrary dimension and degree of exactness with strictly positive weights is known.

After an overview of the basic theory of multidimensional numerical integration, two methods to construct cubature rules will be discussed in detail. This is, on the one hand, to the construction by means of moment equations, and secondly, the Smolyak algorithm. In the last part of the chapter the so-called compound rules are introduced, which are based on domain-decomposition.

3.2.1 Stability Factor

The stability of a cubature rule is a crucial issue and strongly dependent on the number and magnitude of negative weights used by the cubature rule. The influence of negative weights can have an unfavourable effect on the integration process and should therefore be avoided or at least be minimized. Especially in filtering applications, an unstable cubature rule can lead to serious problems. Due to the iterative structure of the filtering process, rounding errors can accumulate which may easily lead to inaccurate results, or even to the divergence of the filter. This may in particular arise, if system and/or measurement equation are highly nonlinear.

Evaluating a function a standard measure to quantify the stability of a cubature (and also quadrature) rule can be motivated on the basis of rounding errors. The abscissae χ_l are d-dimensional and of the form

$$\chi_l = \begin{bmatrix} \chi_{l,1} & \chi_{l,2} & \cdots & \chi_{l,d} \end{bmatrix}', \; l = 1, \ldots, n. \quad (3.76)$$

The biased function value at the point χ_l shall be denoted as $\tilde{f}(\chi_l)$. Defective values can arise, for example, due to rounding errors during the evaluation or in cases, where the function values are only given in tabulated form. The absolute difference

---
[4]http://nines.cs.kuleuven.be//research/ecf.

3.2 Multidimensional Deterministic Numerical Integration

between the biased and unbiased function value is given by

$$\left| \tilde{f}(\chi_l) - f(\chi_l) \right| \leq \epsilon. \tag{3.77}$$

This results in the following estimate for the rounding error of the cubature rule:

$$\left| \sum_{l=1}^{n} \tilde{f}(\chi_l) \alpha_l - \sum_{l=1}^{n} f(\chi_l) \alpha_l \right| = \left| \sum_{l=1}^{n} (\tilde{f}(\chi_l) - f(\chi_l)) \alpha_l \right| \leq \epsilon \sum_{l=1}^{n} |\alpha_l|. \tag{3.78}$$

If all weights are positive, the rounding error is therefore bounded by

$$\epsilon \sum_{l=1}^{n} |\alpha_l| = \epsilon \sum_{l=1}^{n} \alpha_l = \epsilon \int_D w(x) \, dx. \tag{3.79}$$

In the case where some of the weights are negative, the error bound increases, because then

$$\epsilon \sum_{l=1}^{n} |\alpha_l| > \epsilon \sum_{l=1}^{n} \alpha_l = \epsilon \int_D w(x) \, dx. \tag{3.80}$$

From the previous statements, the stability measure SF is derived (cf. Antia 1995, pp. 192):

$$\text{SF} = \frac{\sum_{l=1}^{n} |\alpha_l|}{\int_D w(x) \, dx} = \frac{\sum_{l=1}^{n} |\alpha_l|}{\sum_{l=1}^{n} \alpha_l}. \tag{3.81}$$

For the stability factor it holds that $\text{SF} \geq 1$. A cubature rule with all-positive weights has an optimal stability factor of $\text{SF} = 1$. It must be noted that SF is a *theoretical* measure of stability. That means that the use of negative weights does not generally lead to poor results and the performance of a cubature rule is influenced by many factors which define the characteristics of the considered integral. Nevertheless, on the whole, negative weights lead to adverse effects and should therefore be avoided.

3.2.2 A Lower Bound for the Number of Abscissae

Möller's lower bound (Möller 1979) is the sharpest bound known for the cases where

$$\mathcal{I}\left[x_1^{a_1} x_2^{a_2} \ldots x_d^{a_d} \cdot w(x)\right] = \int_D x_1^{a_1} x_2^{a_2} \ldots x_d^{a_d} w(x) \, dx \tag{3.82}$$

is centrally symmetric. The integral $\mathcal{I}\left[x_1^{a_1} x_2^{a_2} \ldots x_d^{a_d} \cdot w(x)\right]$ is called centrally symmetric if it vanishes, whenever the sum $\sum_{l=1}^{d} a_l$ is odd (cf. Möller 1976, p. 186). For a cubature rule with a degree of exactness of $m = 2k - 1$ the lower bound is

$$n_{\min} = \begin{cases} \binom{d+k-1}{d} + \sum_{l=1}^{d-1} 2^{l-d}\binom{l+k-1}{l}, & \text{k even} \\ \binom{d+k-1}{d} + \sum_{l=1}^{d-1} \left(1 - 2^{l-d}\right)\binom{l+k-2}{l}, & \text{k odd} \end{cases} \tag{3.83}$$

It is important to keep in mind that there is no guarantee that the lower bound can be attained when searching for an efficient cubature rule. Therefore, it can only serve as an indication of how many abscissae are *at least* needed.

3.2.3 Polynomials in d Dimensions

Multidimensional polynomials are linear combinations of multidimensional monomials. The total degree of a multidimensional monomial is defined as the sum of its exponents. Analogous to the one-dimensional case, the degree of multidimensional polynomial is defined as the highest total degree of its multidimensional monomials. As an example, the two-dimensional polynomial

$$p(x_1, x_2) = 2 + x_1 + x_1 x_2 \tag{3.84}$$

is of degree $m = 2$, because the $x_1 x_2$ is of total degree $1 + 1 = 2$ which is higher than the total degree of the other two multidimensional monomials. The space of *complete* two-dimensional polynomials of degree $m = 2$ is a six-dimensional vector space, which is spanned by

$$\left\{1, \ x_1, \ x_1^2, \ x_2, \ x_2^2, \ x_1 x_2\right\}. \tag{3.85}$$

Therefore, the complete d-dimensional polynomials of degree m constitute a $\binom{m+d}{m}$-dimensional vector space (cf. Jetter 2006, p. 199), spanned by the basis

$$\left\{x_1^{a_1} x_2^{a_2} \ldots x_d^{a_d} \,\big|\, a_1, a_1, \ldots a_d \in \mathbb{N}_0, \ \sum_{i=1}^{d} a_i \leq m\right\}. \tag{3.86}$$

The supplement "complete" means that the mentioned polynomials are constructed by all possible (multidimensional) monomials for which (3.86) holds.

3.2.4 Product Cubature Rules

A cubature rule which is exact for d-dimensional polynomials of degree m can be constructed by extending a quadrature rule, which is exact for one-dimensional polynomials of degree m, to a d-dimensional tensor product. For this approach it is crucial that the assumption of factorizability (3.2) of the weight function is satisfied. The integral can then be written as:

$$\mathcal{I}[fw] = \int_D f(x) w(x) \, dx$$

$$= \int_{I_1} \cdots \left(\int_{I_{d-1}} \left(\int_{I_d} f(x) w(x_d) \, dx_d \right) w(x_{d-1}) \, dx_{d-1} \right) \cdots w(x_1) \, dx_1 \tag{3.87}$$

The dimension-wise integration can be performed by a quadrature rule, which leads to the approximation:

$$\mathcal{I}[fw] \approx \sum_{l_1=1}^{n} \sum_{l_2=1}^{n} \cdots \sum_{l_d=1}^{n} f(\chi_{l_1}, \ldots, \chi_{l_d}) \, \vartheta_{l_1} \cdot \ldots \cdot \vartheta_{l_d} \tag{3.88}$$

$$= \sum_{l=1}^{n^d} f(\boldsymbol{\chi}_l) \, \alpha_l. \tag{3.89}$$

Thus, the multidimensional integral can be approximated by the tensor product of a quadrature rule of choice:

$$\mathcal{I}[fw] \approx (Q_1 \otimes Q_2 \otimes \cdots \otimes Q_d) [fw], \text{ with } Q_1 = Q_2 = \ldots = Q_d. \tag{3.90}$$

The resulting rules are often called product cubature rules (cf. Engels 1980, pp. 74–76). As a consequence, the number of abscissae at which the function has to be evaluated rises exponentially. The evolution of a tensor product in dependence on the dimension d is shown in Fig. 3.1. Because of the rapid growth of the number of abscissae, product cubature rules are not suitable for application to high-dimensional problems.

A look at the integral of the weighted multivariate polynomial $p(x) = \left(x_1^m x_2^m \cdots x_d^m\right) w(x)$ reveals the degree of polynomial exactness provided by the tensor product approach. The integral can be written as

$$\mathcal{I}[pw] = \int_{I_1} x_1^m w(x_1) \, dx_1 \int_{I_2} x_2^m w(x_2) \, dx_2 \ldots \int_{I_d} x_d^m w(x_d) \, dx_d. \tag{3.91}$$

Each one-dimensional integral involved in the above product can be exactly evaluated by a quadrature rule of exactness m. It follows that the tensor product

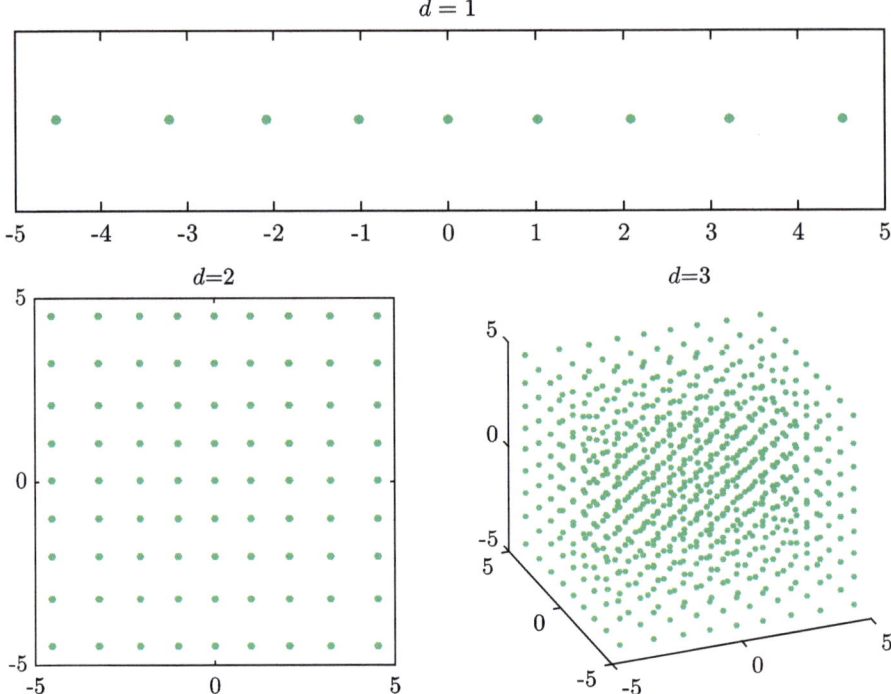

Fig. 3.1 Tensor products

approach is exact for complete multivariate polynomials of degree m and beyond this, for a special class of multivariate polynomials, where each one-dimensional monomial is of degree $\leq m$. This reveals, the product cubature rules are based on the tensor product of one-dimensional polynomials. As an example, for $d = 2$ and $m = 2$ this results in the constitution of the basis

$$\{1, x_1, x_1^2\} \otimes \{1, x_2, x_2^2,\} = \{1, x_1, x_1^2, x_2, x_2^2, x_1 x_2, x_1^2 x_2, x_2^2 x_1, x_1^2 x_2^2\}. \tag{3.92}$$

As can be seen, multidimensional monomials of degree three and four may occur in the polynomials which stem from the vector space spanned by the above basis. But nevertheless, the basis is only complete of total degree two, because, for example, x_1^3 and x_1^4 are not part of the basis. With respect to smooth functions, the Taylor series expansion $T_2[f(x)]$ about the point $\boldsymbol{a} = (a_1, a_2)$ gives an insight to the question why the use of the tensor product basis is inefficient. For a sufficiently smooth

3.2 Multidimensional Deterministic Numerical Integration

two-dimensional function the Taylor series expansion to the second order is

$$T_2\left[f(x)\right] = f(a) + \frac{\partial f(a)}{\partial x_1}(x_1 - a_1) + \frac{\partial f(a)}{\partial x_2}(x_2 - a_2)$$
$$+ \frac{\partial^2 f(a)}{\partial x_1 \partial x_2}(x_1 - a_1)(x_2 - a_2) \qquad (3.93)$$
$$+ \frac{1}{2}\frac{\partial^2 f(a)}{\partial x_1 \partial x_1}(x_1 - a_1)^2 + \frac{1}{2}\frac{\partial^2 f(a)}{\partial x_2 \partial x_2}(x_2 - a_2)^2.$$

After expanding the product terms and collecting all constants it becomes clear that the Taylor series expansion up to the second order yields a complete multidimensional polynomial of degree $m = 2$:

$$T_2\left[f(x)\right] = c_1 + c_2 x_1 + c_3 x_2 + c_4 x_1 x_2 + c_5 x_1^2 + c_6 x_2^2. \qquad (3.94)$$

Generally speaking, around a the Taylor series expansion of order m asymptotically yields a convergence of degree m. Although the tensor basis contains much more elements, no higher order of convergence can be achieved. This is due to the fact that not all multidimensional monomials of total degree $m + 1$ are included. So approximations by complete polynomials asymptotically are as good as those of the tensor product but with far fewer abscissae (cf. Judd 1998, pp. 239–240). A basic task in the construction of efficient cubature rules is therefore to limit the degree of exactness to complete polynomials of degree m.

3.2.5 Moment Equations for the d-Dimensional Case

As in the one-dimensional case, cubature rules can also be constructed using the moment equations. In the further course therefore the multidimensional moment equations are derived. M shall be the matrix whose columns are the d-dimensional abscissae χ_l:

$$M = \begin{bmatrix} \chi_{1,1} & \chi_{1,2} & \cdots & \chi_{1,n} \\ \chi_{2,1} & \chi_{2,2} & \cdots & \chi_{2,n} \\ \chi_{3,1} & \chi_{3,2} & \cdots & \chi_{3,n} \\ \vdots & \vdots & \vdots & \vdots \\ \chi_{d,1} & \chi_{d,2} & \cdots & \chi_{d,n} \end{bmatrix}. \qquad (3.95)$$

The matrix A of the system of equations $A\alpha = b$ comprises all combinations of the rows of M, here referred to as M_i, $i = 1 \ldots d$, of the form

$$\left\{ M_1^{a_1} \odot M_2^{a_2} \odot \ldots \odot M_d^{a_d} \middle| a_1, a_2, \ldots a_d \in \mathbb{N}_0, \sum_{i=1}^{d} a_i \leq m \right\}.^5 \qquad (3.96)$$

In the vector b, the integrals of the corresponding weighted monomials are contained (3.86). As an example, the moment equations for $d = 2$ and polynomial degree of exactness $m = 2$ read

$$\underbrace{\begin{bmatrix} 1 & 1 & \ldots & 1 \\ \chi_{1,1} & \chi_{1,2} & \ldots & \chi_{1,n} \\ \chi_{1,1}^2 & \chi_{1,2}^2 & \ldots & \chi_{1,n}^2 \\ \chi_{2,1} & \chi_{2,2} & \ldots & \chi_{2,n} \\ \chi_{2,1}^2 & \chi_{2,2}^2 & \ldots & \chi_{2,n}^2 \\ \chi_{1,1}\chi_{2,1} & \chi_{1,2}\chi_{2,2} & \ldots & \chi_{1,n}\chi_{2,n} \end{bmatrix}}_{A} \cdot \underbrace{\begin{bmatrix} \alpha_1 \\ \vdots \\ \vdots \\ \vdots \\ \vdots \\ \alpha_n \end{bmatrix}}_{\alpha} = \underbrace{\begin{bmatrix} \int_D w(x) \, dx \\ \int_D x_1 \cdot w(x) \, dx \\ \int_D x_1^2 \cdot w(x) \, dx \\ \int_D x_2 \cdot w(x) \, dx \\ \int_D x_2^2 \cdot w(x) \, dx \\ \int_D x_1 x_2 \cdot w(x) \, dx \end{bmatrix}}_{b}. \qquad (3.97)$$

In order to construct a cubature rule, the abscissae χ_l and weights α_l have to be found which satisfy the above equations for given d, m and n. The arising optimization problem therefore is nonlinear with respect to χ_l. To maximize the efficiency of the cubature rule, it is the goal to decrease n as far as possible. For increasing d, this results in highly overdetermined systems.

Simplification of the d-Dimensional Moment Equations As for the one-dimensional case, the moment equations for the d-dimensional case can be simplified by using the property of central symmetry. In the multidimensional case

$$\int_D x_1^{a_1} x_2^{a_2} \ldots x_d^{a_d} w(x) \, dx \qquad (3.98)$$

is called centrally symmetric if it vanishes, whenever the sum $a_1 + a_2 + \ldots + a_d$ is odd (cf. Möller 1976, p. 186). Therefore the following approach can, inter alia, be applied in the case of the weight function $w(x) = 1$ with the domain $[-1, 1]^d$ and

---

[5]

$$M_i := M_{i,j}, \ j = 1 \ldots n$$

$$[a \ b]^p := [a^p \ b^p] \ \text{(Element-wise exponential)}$$

$$[a \ b] \odot [c \ d] := [ac \ bd] \ \text{(Element-wise product)}.$$

3.2 Multidimensional Deterministic Numerical Integration

the Gaussian weight with the domain $\mathbb{R}^d$. Analogous to the remarks in Sect. 3.1.2 the procedure is described by the example of the latter weight function, namely the multivariate standard normal distribution, $\frac{1}{\sqrt{(2\pi)^d}} e^{-\frac{x'x}{2}}$. Because $\int_{-\infty}^{\infty} x^j \frac{1}{\sqrt{2\pi}} e^{-\frac{x^2}{2}} dx = 0$ for $j = 2k+1$, $k = 0, 1, 2 \ldots$ it follows that

$$\int_{-\infty}^{\infty} x_1^{a_1} x_2^{a_2} \ldots x_d^{a_d} \frac{1}{\sqrt{(2\pi)^d}} e^{-\frac{x'x}{2}} dx = \mathbb{E}\left[x_1^{a_1} x_2^{a_2} \ldots x_d^{a_d}\right] = 0, \tag{3.99}$$

whenever at least one of the exponents a_i is odd. This includes the cases where the sum of the exponents a_i is odd. Thus, all equations of the system $A\alpha = b$,

$$\left[M_1^{a_1} \odot M_2^{a_2} \odot \ldots \odot M_d^{a_d}\right] \cdot \alpha = \mathbb{E}\left[x_1^{a_1} x_2^{a_2} \ldots x_d^{a_d}\right], \quad \sum_{i=1}^{d} a_i \leq m, \tag{3.100}$$

are equal to zero, if $\sum_{i=1}^{d} a_i$ is odd.

Both for an odd and for an even number of abscissae this property can be used to decrease the number of equations, analogous to the one-dimensional case. In the following, the approach will be described in detail for the case of an even number of abscissae. For this case, the matrix M of restricted abscissae reads

$$M = \begin{bmatrix} \chi_{1,1} & \chi_{1,2} & \cdots & \chi_{1,n/2} & -\chi_{1,n/2} & \cdots & -\chi_{1,2} & -\chi_{1,1} \\ \chi_{2,1} & \chi_{2,2} & \cdots & \chi_{2,n/2} & -\chi_{2,n/2} & \cdots & -\chi_{2,2} & -\chi_{2,1} \\ \chi_{3,1} & \chi_{3,2} & \cdots & \chi_{3,n/2} & -\chi_{3,n/2} & \cdots & -\chi_{3,2} & -\chi_{3,1} \\ \vdots & \vdots & \vdots & \vdots & \vdots & \vdots & \vdots \\ \chi_{d,1} & \chi_{d,2} & \cdots & \chi_{d,n/2} & -\chi_{d,n/2} & \cdots & -\chi_{d,2} & -\chi_{d,1} \end{bmatrix}. \tag{3.101}$$

Hence, it is $\chi_1 = -\chi_n$, $\chi_2 = -\chi_{n-1}, \ldots, \chi_{n/2} = -\chi_{(n/2)+1}$ and the rows M_i therefore have the following structure:

$$M_{i,1} = -M_{i,n}, \, M_{i,2} = -M_{i,n-1}, \ldots, M_{i,(n/2)} = -M_{i,(n/2)+1}. \tag{3.102}$$

Each row M_i is symmetric in terms of its absolute values but asymmetric in terms of its signs. The resulting weights α_l have the structure

$$\alpha_1 = \alpha_n, \, \alpha_2 = \alpha_{n-1}, \ldots, \alpha_{n/2} = \alpha_{(n/2)+1}. \tag{3.103}$$

The weights are symmetric in terms of their absolute values and signs. The restrictions lead to the fact that all equations which relate to results of the form

$$\mathbb{E}\left[x_1^{a_1} x_2^{a_2} \ldots x_d^{a_d}\right], \quad \sum_{i=1}^{d} a_i \text{ is odd,} \tag{3.104}$$

can be deleted, because the restricted abscissae yield the right result, namely zero, anyway (see Appendix D). This leads to an immense relief with respect to the computational effort. To enforce the above restrictions on the weights, the moment equations have to be rewritten. This is best explained by an example, again in the case of the Gaussian weight function. The moment equations for $d = 2$ and $m = 2$ (3.97) are now rewritten, setting $n = 8$ and taking into account the restrictions on the abscissae:

$$\begin{bmatrix} 1 & 1 & \cdots & 1 & 1 & \cdots & 1 \\ \chi_{1,1} & \chi_{1,2} & \cdots & \chi_{1,4} & -\chi_{1,4} & \cdots & -\chi_{1,1} \\ \chi_{1,1}^2 & \chi_{1,2}^2 & \cdots & \chi_{1,4}^2 & \chi_{1,4}^2 & \cdots & \chi_{1,1}^2 \\ \chi_{2,1} & \chi_{2,2} & \cdots & \chi_{2,4} & -\chi_{2,4} & \cdots & -\chi_{2,1} \\ \chi_{2,1}^2 & \chi_{2,2}^2 & \cdots & \chi_{2,4}^2 & \chi_{2,4}^2 & \cdots & \chi_{2,1}^2 \\ \chi_{1,1}\chi_{2,1} & \chi_{1,2}\chi_{2,2} & \cdots & \chi_{1,4}\chi_{2,4} & \chi_{1,4}\chi_{2,4} & \cdots & \chi_{1,1}\chi_{2,1} \end{bmatrix} \cdot \begin{bmatrix} \alpha_1 \\ \alpha_2 \\ \alpha_3 \\ \alpha_4 \\ \alpha_4 \\ \alpha_3 \\ \alpha_2 \\ \alpha_1 \end{bmatrix}$$

$$= \begin{bmatrix} 1 \\ \mathbb{E}[x_1] = 0 \\ \mathbb{E}[x_1^2] = 1 \\ \mathbb{E}[x_2] = 0 \\ \mathbb{E}[x_2^2] = 1 \\ \mathbb{E}[x_1 x_2] = 0 \end{bmatrix}. \tag{3.105}$$

The equations for which the sum of exponents $a_1 + a_2$ is odd can be discarded:

$$\begin{bmatrix} 1 & 1 & \cdots & 1 & 1 & \cdots & 1 \\ \chi_{1,1}^2 & \chi_{1,2}^2 & \cdots & \chi_{1,4}^2 & \chi_{1,4}^2 & \cdots & \chi_{1,1}^2 \\ \chi_{2,1}^2 & \chi_{2,2}^2 & \cdots & \chi_{2,4}^2 & \chi_{2,4}^2 & \cdots & \chi_{2,1}^2 \\ \chi_{1,1}\chi_{2,1} & \chi_{1,2}\chi_{2,2} & \cdots & \chi_{1,4}\chi_{2,4} & \chi_{1,4}\chi_{2,4} & \cdots & \chi_{1,1}\chi_{2,1} \end{bmatrix} \cdot \begin{bmatrix} \alpha_1 \\ \alpha_2 \\ \alpha_3 \\ \alpha_4 \\ \alpha_4 \\ \alpha_3 \\ \alpha_2 \\ \alpha_1 \end{bmatrix}$$

$$= \begin{bmatrix} 1 \\ \mathbb{E}[x_1^2] = 1 \\ \mathbb{E}[x_2^2] = 1 \\ \mathbb{E}[x_1 x_2] = 0 \end{bmatrix}. \tag{3.106}$$

3.2 Multidimensional Deterministic Numerical Integration

Thus, the system of equations gets rewritten into:

$$\begin{bmatrix} 2 & 2 & 2 & 2 \\ 2\chi_{1,1}^2 & 2\chi_{1,2}^2 & 2\chi_{1,3}^2 & 2\chi_{1,4}^2 \\ 2\chi_{2,1}^2 & 2\chi_{2,2}^2 & 2\chi_{2,3}^2 & 2\chi_{2,4}^2 \\ 2\chi_{1,1}\chi_{2,1} & 2\chi_{1,2}\chi_{2,2} & 2\chi_{1,3}\chi_{2,3} & 2\chi_{1,4}\chi_{2,4} \end{bmatrix} \cdot \begin{bmatrix} \alpha_1 \\ \alpha_2 \\ \alpha_3 \\ \alpha_4 \end{bmatrix} = \begin{bmatrix} 1 \\ \mathbb{E}\left[x_1^2\right] = 1 \\ \mathbb{E}\left[x_2^2\right] = 1 \\ \mathbb{E}\left[x_1 x_2\right] = 0 \end{bmatrix}. \quad (3.107)$$

In fact, due to the restrictions on the abscissae and the symmetry of the weight function, the resulting quadrature rule is not only exact for polynomials of degree up to $m = 2$ but also actually for polynomials of degree up to $m = 3$. It must be stressed that in contrast to the above example, equation systems connected to efficient cubature rules in general are not quadratic but highly overdetermined. The goal is to find a solution to the multidimensional moment equations with n being as small as possible.

For the universal case of centrally symmetric integrals and arbitrary choices of d, m and n (even), one can formulate the procedure as follows: The Matrix A^* shall contain all combinations

$$\left\{ M_1^{a_1} \odot M_2^{a_2} \odot \ldots \odot M_d^{a_d} \,\middle|\, a_i \in \mathbb{N}_0, \sum_{i=1}^d a_i \text{ is even}, \sum_{i=1}^d a_i \le m \right\}. \quad (3.108)$$

In the case of even n, the matrix B shall contain the first $n/2$ columns of $2A^*$ and the vector v shall consist of the weights to be calculated, namely $\alpha_1, \alpha_2, \ldots, \alpha_{n/2}$. Furthermore, a vector c is defined which shall contains the corresponding integrals

$$\int_D x_1^{a_1} x_2^{a_2} \ldots x_d^{a_d} w(\boldsymbol{x}) \, d\boldsymbol{x}, \quad \sum_{i=1}^d a_i \text{ is even}, \quad \sum_{i=1}^d a_i \le m. \quad (3.109)$$

For odd n, only minor changes have to be made. In this case the midpoint $\begin{bmatrix} 0 & 0 & \cdots & 0 \end{bmatrix}'$ is introduced into the system of equations and the Matrix M reads

$$M = \begin{bmatrix} \chi_{1,1} & \cdots & \chi_{1,(n-1)/2} & 0 & -\chi_{1,(n-1)/2} & \cdots & -\chi_{1,1} \\ \chi_{2,1} & \cdots & \chi_{2,(n-1)/2} & 0 & -\chi_{2,(n-1)/2} & \cdots & -\chi_{2,1} \\ \chi_{3,1} & \cdots & \chi_{3,(n-1)/2} & 0 & -\chi_{3,(n-1)/2} & \cdots & -\chi_{3,1} \\ \vdots & \vdots & \vdots & \vdots & \vdots & \vdots & \vdots \\ \chi_{d,1} & \cdots & \chi_{d,(n-1)/2} & 0 & -\chi_{d,(n-1)/2} & \cdots & -\chi_{d,1} \end{bmatrix}. \quad (3.110)$$

The matrix B then consists of the first $(n-1)/2$ columns of $2A^*$ and an additional column $\begin{bmatrix} 1 & 0 & \cdots & 0 \end{bmatrix}'$. The vector v contains the weights $\alpha_1, \alpha_2, \ldots, \alpha_{(n+1)/2}$. The first component of this additional column is equal to one, because the midpoint-weight $\alpha_{(n+1)/2}$ only has a multiplicity of one and the components of the final vector of weights, $\boldsymbol{\alpha}$, must fulfill the equation

$$\sum_{l=1}^{n} \alpha_l = 1. \tag{3.111}$$

The resulting simplified multidimensional moment equations in the following will be denoted as:

$$\boldsymbol{Bv} = \boldsymbol{c}. \tag{3.112}$$

Selected Cubature Rules for $m = 3$ and Gaussian Weight In filtering applications, the use of cubature rules with degree of exactness $m = 3$ is very common. To clarify the following concepts, these will be illustrated again for the case $d = 2$. The number of points will be set to $n = 4$ and the abscissae $\boldsymbol{\chi}_l$ and weights α_l shall obey the restrictions which have been described in the previous section. It will be shown that due to a specific choice of abscissae, the system of equations (3.107) can be reduced even further. The abscissae $\boldsymbol{\chi}_l$ have a special structure and are contained in the set

$$\left\{ \begin{bmatrix} \sqrt{2} \\ 0 \end{bmatrix}, \begin{bmatrix} 0 \\ \sqrt{2} \end{bmatrix}, \begin{bmatrix} -\sqrt{2} \\ 0 \end{bmatrix}, \begin{bmatrix} 0 \\ -\sqrt{2} \end{bmatrix} \right\}. \tag{3.113}$$

Additionally to the properties which have been already mentioned, the following holds:

$$\left(\boldsymbol{M}_1^{a_1} \odot \boldsymbol{M}_2^{a_2} \right) \cdot \boldsymbol{\alpha} = 0 \tag{3.114}$$

for all choices of exponents a_1 and a_2 and independent of $\boldsymbol{\alpha}$. This leads to the fact that the fourth equation of (3.107) can be eliminated. In analogy to (3.107), the remaining equations for $n = 4$ then read

$$\begin{bmatrix} 2 & 2 \\ 2 \cdot \left(\sqrt{2} \right)^2 & 2 \cdot 0^2 \\ 2 \cdot 0^2 & 2 \cdot \left(\sqrt{2} \right)^2 \end{bmatrix} \cdot \begin{bmatrix} \alpha_1 \\ \alpha_2 \end{bmatrix} = \begin{bmatrix} 1 \\ \mathbb{E}\left[x_1^2\right] = 1 \\ \mathbb{E}\left[x_2^2\right] = 1 \end{bmatrix} \tag{3.115}$$

and therefore

$$\begin{bmatrix} 2 & 2 \\ 4 & 0 \\ 0 & 4 \end{bmatrix} \cdot \begin{bmatrix} \alpha_1 \\ \alpha_2 \end{bmatrix} = \begin{bmatrix} 1 \\ \mathbb{E}\left[x_1^2\right] = 1 \\ \mathbb{E}\left[x_2^2\right] = 1 \end{bmatrix} \tag{3.116}$$

As to be seen from equation two and three, $\alpha_1 = \alpha_2 = \alpha_3 = \alpha_4 = \frac{1}{4}$. Also the first equation is fulfilled, using these weights. Therefore, also the first equation can be

3.2 Multidimensional Deterministic Numerical Integration

deleted, since the equation is always satisfied anyway. So one arrives at

$$\begin{bmatrix} 4 & 0 \\ 0 & 4 \end{bmatrix} \cdot \begin{bmatrix} \alpha_1 \\ \alpha_2 \end{bmatrix} = \begin{bmatrix} \mathbb{E}\left[x_1^2\right] = 1 \\ \mathbb{E}\left[x_2^2\right] = 1 \end{bmatrix}. \tag{3.117}$$

In conclusion, the presented set of abscissae in combination with the calculated weights satisfy all relevant moment equations required for $d = 2$, $m = 3$ and the Gaussian weight.

These considerations lead to the cubature rule derived by Arasaratnam and Haykin (2009) which they use in their self-developed cubature Kalman filter (CKF). To use this cubature rule for the general case of arbitrary dimension, $n = 2d$ abscissae, forming the following set, are needed:

$$\left\{ \begin{bmatrix} \sqrt{d} \\ 0 \\ \vdots \\ 0 \end{bmatrix}, \begin{bmatrix} 0 \\ \sqrt{d} \\ \vdots \\ 0 \end{bmatrix}, \dots, \begin{bmatrix} 0 \\ 0 \\ \vdots \\ \sqrt{d} \end{bmatrix}, \begin{bmatrix} -\sqrt{d} \\ 0 \\ \vdots \\ 0 \end{bmatrix}, \begin{bmatrix} 0 \\ -\sqrt{d} \\ \vdots \\ 0 \end{bmatrix}, \dots, \begin{bmatrix} 0 \\ 0 \\ \vdots \\ -\sqrt{d} \end{bmatrix} \right\}. \tag{3.118}$$

This yields the moment equations

$$\begin{bmatrix} 2d & 0 & \cdots & 0 \\ 0 & 2d & \cdots & 0 \\ \vdots & 0 & \ddots & \vdots \\ 0 & \cdots & 0 & 2d \end{bmatrix} \cdot \begin{bmatrix} \alpha_1 \\ \vdots \\ \alpha_d \end{bmatrix} = \begin{bmatrix} 1 \\ \vdots \\ 1 \end{bmatrix} \tag{3.119}$$

with the unique solution $\alpha_1 = \alpha_2 = \ldots = \alpha_{2d} = \frac{1}{2d}$.

In early scientific publications, cubature rules often have been derived with respect to the alternative weight function $w(x) = e^{-x^T x}$. Already in 1963, Stroud and Secrest (1963) have published the first cubature rule for the given weight function having a degree of exactness of $m = 3$ and using $2d$ abscissae. The set of abscissae reads

$$\left\{ \begin{bmatrix} \sqrt{\frac{d}{2}} \\ 0 \\ \vdots \\ 0 \end{bmatrix}, \begin{bmatrix} 0 \\ \sqrt{\frac{d}{2}} \\ \vdots \\ 0 \end{bmatrix}, \dots, \begin{bmatrix} 0 \\ 0 \\ \vdots \\ \sqrt{\frac{d}{2}} \end{bmatrix}, \begin{bmatrix} -\sqrt{\frac{d}{2}} \\ 0 \\ \vdots \\ 0 \end{bmatrix}, \begin{bmatrix} 0 \\ -\sqrt{\frac{d}{2}} \\ \vdots \\ 0 \end{bmatrix}, \dots, \begin{bmatrix} 0 \\ 0 \\ \vdots \\ -\sqrt{\frac{d}{2}} \end{bmatrix} \right\}. \tag{3.120}$$

The weights are to be calculated by using the formula

$$\alpha_l = \frac{1}{2d} \cdot V \tag{3.121}$$

with

$$V = \int_D e^{-x'x} dx = \pi^{d/2}. \qquad (3.122)$$

Obviously, by the change of variables $y = \frac{x}{\sqrt{2}}$ it shows that

$$\int_D f(y) e^{-y'y} dy = \frac{1}{\sqrt{2}^d} \int_D f\left(\frac{x}{\sqrt{2}}\right) e^{-\frac{x'x}{2}} dx. \qquad (3.123)$$

Thus, the cubature rule of Stroud and Secrest can be easily transformed and used in order to approximate Gaussian integrals. Comparing the performed change of variables and the set of abscissae (3.118) it becomes clear that the transformed abscissae are equal to those of Arasaratnam and Haykin (2009). The same holds for the transformed weights. Multiplying the standardization factor of the Gaussian distribution, $\frac{1}{\sqrt{(2\pi)^d}}$, and the reciprocal of the Jacobi determinant $\frac{1}{\sqrt{2}^d}$, leads to

$$\alpha_l = \frac{1}{\sqrt{(2\pi)^d}} \cdot \sqrt{2}^d \cdot \frac{1}{2d} \cdot \pi^{d/2} = \frac{1}{2d}, \; l = 1, 2, \ldots, 2d. \qquad (3.124)$$

Accordingly, the cubature rule of Arasaratnam and Haykin (2009) can be interpreted as the transformation of a cubature rule which is already well known from the relevant literature.

Another important cubature rule widely used in filter applications is the unscented transform of Julier et al. (1995). It has already been described in Sect. 2.3.2 but will now be derived in a different way. The unscented transform in connection with the unscented Kalman filter was the first ever published approach to filtering by applying a cubature rule. In the following it will turn out that the cubature rule of Arasaratnam and Haykin (2009) is a special case of the unscented transform due to the fact that the UKF equals the CKF, if $\kappa = 0$. A close look at (3.117) shows that also other sets of abscissae could be chosen which meet the restrictions (3.102) and (3.114):

$$\left\{ \begin{bmatrix} \sqrt{2+\kappa} \\ 0 \end{bmatrix}, \begin{bmatrix} 0 \\ \sqrt{2+\kappa} \end{bmatrix}, \begin{bmatrix} -\sqrt{2+\kappa} \\ 0 \end{bmatrix}, \begin{bmatrix} 0 \\ -\sqrt{2+\kappa} \end{bmatrix} \right\}. \qquad (3.125)$$

The use of the parameter κ opens up the possibility to set values for higher order monomials, for example, x_1^4 and x_2^4. To ensure that the weights add up to one, an additional point in form of the midpoint $[0\ 0]'$ is introduced. To meet the restriction (3.103) and because the number of abscissae and weights is now odd,

3.2 Multidimensional Deterministic Numerical Integration

the resulting system of equations reads

$$\begin{bmatrix} 2 \cdot (\sqrt{2+\kappa})^2 & 2 & 1 \\ 2 \cdot 0^2 & 2 \cdot 0^2 & 0 \\ 2 \cdot 0^2 & 2 \cdot (\sqrt{2+\kappa})^2 & 0 \end{bmatrix} \cdot \begin{bmatrix} \alpha_1 \\ \alpha_2 \\ \alpha_3 \end{bmatrix} = \begin{bmatrix} 1 \\ \mathbb{E}[x_1^2] = 1 \\ \mathbb{E}[x_2^2] = 1 \end{bmatrix}. \quad (3.126)$$

So one arrives at

$$\begin{bmatrix} 2 & 2 & 1 \\ 4+2\kappa & 0 & 0 \\ 0 & 4+2\kappa & 0 \end{bmatrix} \cdot \begin{bmatrix} \alpha_1 \\ \alpha_2 \\ \alpha_3 \end{bmatrix} = \begin{bmatrix} 1 \\ \mathbb{E}[x_1^2] = 1 \\ \mathbb{E}[x_2^2] = 1 \end{bmatrix} \quad (3.127)$$

with the unique solution

$$\alpha_1 = \alpha_2 = \alpha_4 = \alpha_5 = \frac{1}{4+2\kappa},$$

$$\alpha_3 = 1 - \left(\sum_{i=1}^{2} \alpha_i + \sum_{i=4}^{5} \alpha_i \right) = \frac{\kappa}{2+\kappa}. \quad (3.128)$$

Obviously, for $\kappa = 0$ the equation system (3.127) yields the same result as (3.117) and therefore the abscissae and weights then equal those of the cubature rule of Arasaratnam and Haykin (2009). Furthermore, the parameter κ can be determined in a way, so that, for example, the abscissae and weights integrate the Gaussian integral $\int_{-\infty}^{\infty} x^4 \frac{1}{\sqrt{2\pi}} e^{-\frac{x^2}{2}} dx = \mathbb{E}[x^4] = 3$ exactly.[6] For this purpose one has to solve the following system of equations for κ:

$$\begin{bmatrix} (\sqrt{2+\kappa})^4 & 0 & (-\sqrt{2+\kappa})^4 & 0 & 0 \\ 0 & (\sqrt{2+\kappa})^4 & 0 & (-\sqrt{2+\kappa})^4 & 0 \end{bmatrix} \cdot \begin{bmatrix} \alpha_1 \\ \alpha_2 \\ \alpha_3 \\ \alpha_4 \\ \alpha_5 \end{bmatrix} = \begin{bmatrix} \mathbb{E}[x_1^4] = 3 \\ \mathbb{E}[x_2^4] = 3 \end{bmatrix}$$

$$\Rightarrow \kappa = 1. \quad (3.129)$$

---

[6] Any other one-dimensional moment $\mathbb{E}[x^c]$ would be possible, too.

For the d-dimensional case, the set of abscissae reads

$$\left\{ \begin{bmatrix} \sqrt{d+\kappa} \\ 0 \\ \vdots \\ 0 \end{bmatrix}, \ldots, \begin{bmatrix} 0 \\ 0 \\ \vdots \\ \sqrt{d+\kappa} \end{bmatrix}, \begin{bmatrix} 0 \\ 0 \\ \vdots \\ 0 \end{bmatrix}, \begin{bmatrix} -\sqrt{d+\kappa} \\ 0 \\ \vdots \\ 0 \end{bmatrix}, \ldots, \begin{bmatrix} 0 \\ 0 \\ \vdots \\ -\sqrt{d+\kappa} \end{bmatrix} \right\}. \tag{3.130}$$

This yields the moment equations

$$\begin{bmatrix} 2 & 2 & \cdots & 2 & 1 \\ 2d+2\kappa & 0 & \cdots & 0 & 0 \\ 0 & 2d+2\kappa & \cdots & 0 & 0 \\ \vdots & 0 & \ddots & \vdots & 0 \\ 0 & \cdots & 0 & 2d+2\kappa & 0 \end{bmatrix} \cdot \begin{bmatrix} \alpha_1 \\ \vdots \\ \vdots \\ \alpha_{d+1} \end{bmatrix} = \begin{bmatrix} 1 \\ \vdots \\ \vdots \\ 1 \end{bmatrix} \tag{3.131}$$

with the unique solution $\alpha_{(2d+2)/2} = \frac{k}{d+\kappa}$ and $\frac{1}{2(d+\kappa)}$ for the remaining weights α_l. For the d-dimensional case, in order to secure the exactness of the cubature rule for the Gaussian moment $\mathbb{E}\left[x^4\right] = 3$, one has to solve the equation

$$2 \cdot \left(\sqrt{d+k}\right)^4 \cdot \frac{1}{2(d+k)} = 3 \tag{3.132}$$

for κ which yields $\kappa = 3-d$. According to this, for $d > 3$ the occurrence of negative weights then is unavoidable. It has to be kept in mind that the cth moment of the d-dimensional (standard) normal distribution is a tensor consisting of all moments

$$\mathbb{E}\left[x_1^{a_1} x_2^{a_2} \cdots x_d^{a_d}\right], \text{ with } \sum_{i=1}^{d} a_i = c. \tag{3.133}$$

Thus, the possibility to integrate a single higher moment, for example, the fourth moment $\mathbb{E}\left[x^4\right]$, will not lead to a noteworthy improvement of the cubature rule because all other fourth moments still cannot be integrated correctly. So as previously indicated, the parameter κ seems more or less useless.

3.2.6 Smolyak Cubature

The technique of Smolyak uses tensor products of quadrature rules as building blocks for the construction of cubature rules. In contrast to the ordinary product rules described in Sect. 3.2.4, Smolyak cubature rules are very efficient in terms of

3.2 Multidimensional Deterministic Numerical Integration

the number of used abscissae. Furthermore, due to the compactness of the algorithm, rules for high dimensions and of high exactness can be computed with small effort.

The starting point is a series of quadrature rules Q_i, $i = 1, 2, \ldots$. These rules are supposed to possess degrees of exactness so that $m_i < m_{i+1}$. Using a telescoping sum, with $Q_0[fw] = 0$,

$$Q_j[fw] = \sum_{i=1}^{j} (Q_i - Q_{i-1})[fw]. \tag{3.134}$$

Dropping $[fw]$ for the sake of a better readability, the tensor product of the quadrature rule Q_j can therefore alternatively be written as

$$Q_{j_1} \otimes Q_{j_2} \otimes \cdots \otimes Q_{j_d} = \bigotimes_{k=1}^{d} \sum_{i_k=1}^{j} (Q_{i_k} - Q_{i_k-1}) = \sum_{i_1,\ldots,i_d=1}^{j} \bigotimes_{k=1}^{d} (Q_{i_k} - Q_{i_k-1}). \tag{3.135}$$

Because the quadrature rules are ordered by ascending exactness it follows that

$$\sum_{i_1,\ldots,i_d=1}^{\infty} \bigotimes_{k=1}^{d} (Q_{i_k} - Q_{i_k-1}) \tag{3.136}$$

equals the exact value of the integral which has to be evaluated. The basic idea of the Smolyak cubature (cf. Smolyak 1963; Sickel and Ullrich 2006, pp. 1–2) is to approximate

$$\sum_{i_1,\ldots,i_d=1}^{\infty} \bigotimes_{k=1}^{d} (Q_{i_k} - Q_{i_k-1}) \tag{3.137}$$

by a truncated tensor product of the form

$$\sum_{i_1+\cdots+i_d \leq q} \bigotimes_{k=1}^{d} (Q_{i_k} - Q_{i_k-1}), \quad q \geq d. \tag{3.138}$$

In the following, the benefits arising from this kind of approximation as well as its algorithmical implementation will be described.

Derivation of the Smolyak Algorithm First, the difference of two quadrature rules is defined as $\Delta_0 = 0$, $\Delta_1 = Q_1$, $\Delta_i = Q_i - Q_{i-1}$. Setting $\boldsymbol{i} = (i_1, i_2, \ldots, i_d)$, the Smolyak algorithm can be formulated as:

$$A(q, d) = \sum_{\|\boldsymbol{i}\|_1 \leq q} \Delta_{i_1} \otimes \Delta_{i_2} \otimes \cdots \otimes \Delta_{i_d} = \sum_{\|\boldsymbol{i}\|_1 \leq q} \bigotimes_{k=1}^{d} \Delta_{i_k}, \quad q \geq d. \tag{3.139}$$

Example: $q = 4$, $d = 3$ yields the index-vectors $(1, 1, 1), (1, 1, 2), (1, 2, 1),$ $(2, 1, 1), i \in \mathbb{N}^3$

$$\begin{aligned} A(4,3) &= (\Delta_1 \otimes \Delta_1 \otimes \Delta_1) + (\Delta_1 \otimes \Delta_1 \otimes \Delta_2) \\ &+ (\Delta_1 \otimes \Delta_2 \otimes \Delta_1) + (\Delta_2 \otimes \Delta_1 \otimes \Delta_1) \\ &= (Q_1 \otimes Q_1 \otimes Q_1) + (Q_1 \otimes Q_1 \otimes (Q_2 - Q_1)) \\ &+ (Q_1 \otimes (Q_2 - Q_1) \otimes Q_1) + ((Q_2 - Q_1) \otimes Q_1 \otimes Q_1) \,. \end{aligned} \quad (3.140)$$

A further expansion leads to:

$$\begin{aligned} A(4,3) = &\underbrace{(Q_1 \otimes Q_1 \otimes Q_2)}_{\gamma=(1,1,2)} + \underbrace{(Q_1 \otimes Q_2 \otimes Q_1)}_{\gamma=(1,2,1)} \\ &+ \underbrace{(Q_2 \otimes Q_1 \otimes Q_1)}_{\gamma=(2,1,1)} - 2\underbrace{(Q_1 \otimes Q_1 \otimes Q_1)}_{\gamma=(1,1,1)} \,. \end{aligned} \quad (3.141)$$

Obviously, $A(q, d)$ consists of all possible tensor products of the form

$$\left(Q_{\gamma_1} \otimes Q_{\gamma_2} \otimes \ldots \otimes Q_{\gamma_d}\right), \; \|\boldsymbol{\gamma}\|_1 \leq q. \quad (3.142)$$

Each index-vector $\boldsymbol{\gamma}$ can occur several times, with different signs. The task is now to identify the multiplicity of each index-vector $\boldsymbol{\gamma}$ and its affiliated signs. This can be achieved by tracing each single $\boldsymbol{\gamma}$ back to the possible index-vectors $\boldsymbol{i}$ which it stems from.

Example: $\boldsymbol{\gamma} = (1, 1, 1)$, therefore $(Q_1 \otimes Q_1 \otimes Q_1)$. From which index-vectors $\boldsymbol{i}$ can $\boldsymbol{\gamma}$ stem from? The following options are possible:

$$\boldsymbol{i} = (1,1,1) \Rightarrow ((Q_1 - Q_0) \otimes (Q_1 - Q_0) \otimes (Q_1 - Q_0)) \text{ and } \boldsymbol{\gamma} + \underbrace{(0,0,0)}_{\eta} = \boldsymbol{i}$$

$$\boldsymbol{i} = (2,1,1) \Rightarrow ((Q_2 - Q_1) \otimes (Q_1 - Q_0) \otimes (Q_1 - Q_0)) \text{ and } \boldsymbol{\gamma} + \underbrace{(1,0,0)}_{\eta} = \boldsymbol{i}$$

$$\boldsymbol{i} = (1,2,1) \Rightarrow ((Q_1 - Q_0) \otimes (Q_2 - Q_1) \otimes (Q_1 - Q_0)) \text{ and } \boldsymbol{\gamma} + \underbrace{(0,1,0)}_{\eta} = \boldsymbol{i}$$

$$\boldsymbol{i} = (1,1,2) \Rightarrow ((Q_1 - Q_0) \otimes (Q_1 - Q_0) \otimes (Q_2 - Q_1)) \text{ and } \boldsymbol{\gamma} + \underbrace{(0,0,1)}_{\eta} = \boldsymbol{i}$$

(3.143)

3.2 Multidimensional Deterministic Numerical Integration

For the case $q > 4$, for instance, $q = 6$, also other options would have to be taken into account:

$$i = (2,2,2) \Rightarrow ((Q_2 - Q_1) \otimes (Q_2 - Q_1) \otimes (Q_2 - Q_1)) \text{ and } \gamma + \underbrace{(1,1,1)}_{\eta} = i. \quad (3.144)$$

The index-vectors η serve as indicators. For example, the index-vector $i = (2, 1, 1)$, shown above, leads to the tensor product

$$((Q_2 - Q_1) \otimes (Q_1 - Q_0) \otimes (Q_1 - Q_0)). \quad (3.145)$$

After multiplying out the terms, the tensor product of interest, $-(Q_1 \otimes Q_1 \otimes Q_1)$, arises. The sign is negative, because Q_1 has a negative sign in the first bracket of (3.145). This is indicated by the vector $\eta = (1,0,0)$. Furthermore, it should be noted that, due to the structure of the tensor product (3.145), it is $i = \gamma + \eta$.

Concluding it can be stated that, if $\|\eta\|_1$ is odd, $\left(Q_{\gamma_1} \otimes Q_{\gamma_1} \otimes \ldots \otimes Q_{\gamma_d}\right)$ has a negative sign. Adding all possible combinations of $\eta \in \{0,1\}^d$ to a certain γ leads to the index-vectors i where this γ stems from. Summarizing, for every γ with $\|\gamma\|_1 < q$ one has to find all η, so that $\|\gamma + \eta\|_1 < q \Leftrightarrow \|\eta\|_1 \leq q - \|\gamma\|_1$. The number of possible index-vectors η is equal to the multiplicity of the index-vector γ and the norms $\|\eta\|_1$ determine the associated signs. The previous considerations lead to the following representation:

$$A(q,d) = \sum_{\|\gamma\|_1 \leq q} \sum_{\substack{\eta \in \{0,1\}^d \\ \|\eta\|_1 \leq q - \|\gamma\|_1}} (-1)^{|\eta|} \bigotimes_{k=1}^{d} Q_{\gamma_k}, \ q \geq d. \quad (3.146)$$

The sum

$$\sum_{\substack{\eta \in \{0,1\}^d \\ \|\eta\|_1 \leq q - \|\gamma\|_1}} (-1)^{\|\eta\|_1} = \sum_{j=0}^{q-\|\gamma\|_1} -1^j \binom{d}{j} \quad (3.147)$$

can be reformulated into (cf. Gradshtein et al. 2007)

$$\sum_{j=0}^{q-\|\gamma\|_1} (-1)^j \binom{d}{j} = (-1)^{q-\|\gamma\|_1} \binom{d-1}{q - \|\gamma\|_1}. \quad (3.148)$$

The final form of the Smolyak cubature rule can now be stated as (cf. Wasilkowski and Wozniakowski 1995):

$$A(q,d) = \sum_{\|\boldsymbol{\gamma}\|_1 \leq q} (-1)^{q-\|\boldsymbol{\gamma}\|_1} \binom{d-1}{q-\|\boldsymbol{\gamma}\|_1} \bigotimes_{k=1}^{d} Q_{\gamma_k} \qquad (3.149)$$

$$= \sum_{q-d+1 \leq \|\boldsymbol{\gamma}\|_1 \leq q} (-1)^{q-\|\boldsymbol{\gamma}\|_1} \binom{d-1}{q-\|\boldsymbol{\gamma}\|_1} \bigotimes_{k=1}^{d} Q_{\gamma_k}. \qquad (3.150)$$

The last equality holds, because $\binom{n}{k} = 0$ for $k > n$. The following corollary provides a link between the parameters q and d and the polynomial exactness of the Smolyak cubature:

Corollary *If the used quadrature rules Q_i, $i = 1, 2, \ldots$, have a degree of exactness of at least $m_i = 2i - 1$, then $A(q,d)$ has a degree of exactness of at least $m = 2(q-d) + 1$ (cf. Novak and Ritter 1999, p. 507).*

Or in other words: To construct a Smolyak cubature rule of dimension d and with (at least) exactness m, the parameter q must be set to $\left(\frac{m-1}{2}\right) + d$.

As already explained, the vectors $\boldsymbol{\gamma}$ have dimension d and consist of all combinations of the indices $i = 1, 2, \ldots$ (with repetition), for which $\|\boldsymbol{\gamma}\|_1 \leq q$. Therefore the highest possible index is $q - d + 1$. Inserting the expression for q, $q = \left(\frac{m-1}{2}\right) + d$, it turns out that the highest possible index is equal to $\frac{m-1}{2} + 1 = \frac{m+1}{2}$. Thus, the underlying quadrature rules Q_i must have at least the exactness $m_i = 2i - 1$, $i = 1, \ldots, \frac{m+1}{2}$.

As an example, for the construction a Smolyak cubature rule for dimension $d = 5$ and of exactness $m = 7$, the parameter q must be set to $q = \left(\frac{m-1}{2}\right) + d = 8$. The highest index i then is $q - d + 1 = 4$. So it turns out that the underlying quadrature rules, Q_i, $i = 1, 2, 3, 4$, must have *at least* the degrees of exactness $m_1 = 1$, $m_2 = 3$, $m_3 = 5$ and $m_4 = 7$. Independent of d, the most exact underlying quadrature rule must provide at least the same degree of exactness, as the Smolyak cubature rule which is to be constructed.

Some Properties of the Smolyak Cubature In the worst case, the set of abscissae used by $A(q,d)$ is

$$H(q,d) = \bigcup_{q-d+1 \leq \|\boldsymbol{\gamma}\|_1 \leq q} \bigotimes_{k=1}^{d} S_{\gamma_k}. \qquad (3.151)$$

The sets which hold the abscissae of the quadrature rules Q_i, $i = 1, 2, \ldots$ shall be denoted as S_i, $i = 1, 2, \ldots$. If these sets are nested, so that $S_i \subset S_{i+1}$ for $i \geq 1$, the

3.2 Multidimensional Deterministic Numerical Integration

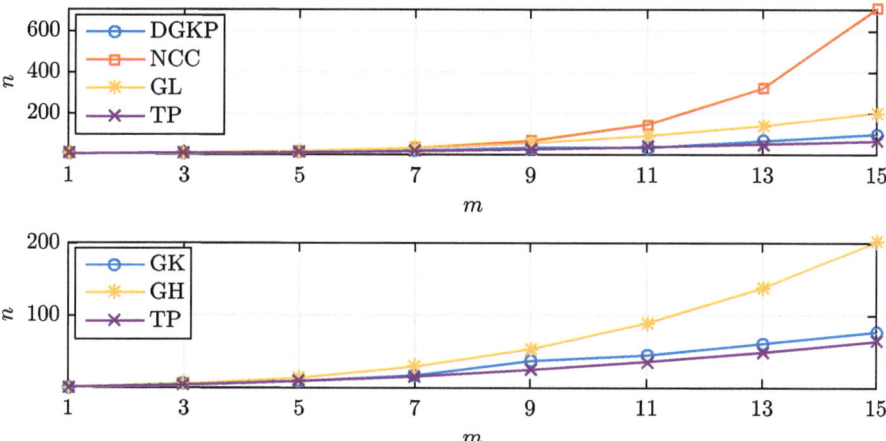

Fig. 3.2 Comparison of various underlying quadrature rules and tensor product for $d = 2$

set $H(q, d)$ changes to (cf. Novak and Ritter 1999, p. 510–511)

$$H(q, d) = \bigcup_{\|\gamma\|_1 = q} \bigotimes_{k=1}^{d} S_{\gamma_k}. \tag{3.152}$$

This implicates that, in many cases, the use of sets which are nested as much as possible are the appropriate choice to reduce the total number of abscissae used by the Smolyak cubature. As the following figures show, the benefit of nested sets is highly dependent on the dimension of the integration problem and on the underlying quadrature rules. The underlying quadrature rules used for the following comparisons have been investigated by Heiss and Winschel (2008), with the only exception being the nested Clenshaw–Curtis rule, which has been studied by Novak and Ritter (1999) in the context of Smolyak cubature. The regions of integration for the unweighted case and the Gaussian weight function are the d-dimensional cube, $[-1, 1]^d$, and $\mathbb{R}^d$, respectively.[7]

Figure 3.2 shows the growth of the abscissae used by the different Smolyak cubature rules in comparison to the tensor product (TP) for the two-dimensional case and the weight function

$$w(x) = 1 \tag{3.153}$$

---

[7] The authors provide downloadable sequences of Smolyak abscissae and weights based on various underlying quadrature rules on their website http://www.sparse-grids.de/. For the unweighted case, the authors use the domain $[0, 1]^d$ which can be easily mapped to $[-1, 1]^d$.

in the upper part and

$$w(x) = \frac{1}{\sqrt{(2\pi)^d}} e^{-\frac{x'x}{2}} dx. \qquad (3.154)$$

in the lower part of the figure. The degrees of exactness vary from $m = 1$ to $m = 15$. With respect to the unweighted case, two nested and one non-nested family of rules have been used, namely the nested Clenshaw–Curtis rules, as described in Sect. 3.1.4, the delayed Gauss–Kronrod–Patterson rules and the Gauss–Legendre rules (Sect. 3.1.3). The delayed Gauss–Kronrod–Patterson rules of Petras (2003) are based on the Patterson approach (Sect. 3.1.3) and are sophisticatedly designed in a way so that the resulting Smolyak cubature rules use a minimal number of abscissae. The arising Smolyak cubature rules are abbreviated with NCC, DGKP and GL. For the computation of the tensor product, also the Gauss–Legendre rules have been used.

In the case of the Gaussian weight function, one nested and one non-nested family of rules have been used. Those are, on the one hand, the quadrature rules of Genz and Keister (Sect. 3.1.3) and, on the other hand, the Gauss–Hermite rules (Sect. 3.1.3). The associated Smolyak cubature rules are abbreviated with GK and GH. Here, the Gauss–Hermite rules have been used as basis for the tensor product.

As can be seen, for low-dimensional integration problems the benefits of using the Smolyak cubature do not emerge. The Smolyak cubature rules, regardless of the underlying quadrature rules, use a higher number of abscissae than the tensor product. For integration problems in higher dimensions the situation is different. Figure 3.3 shows the growth of the number of used abscissae for $d = 10$.

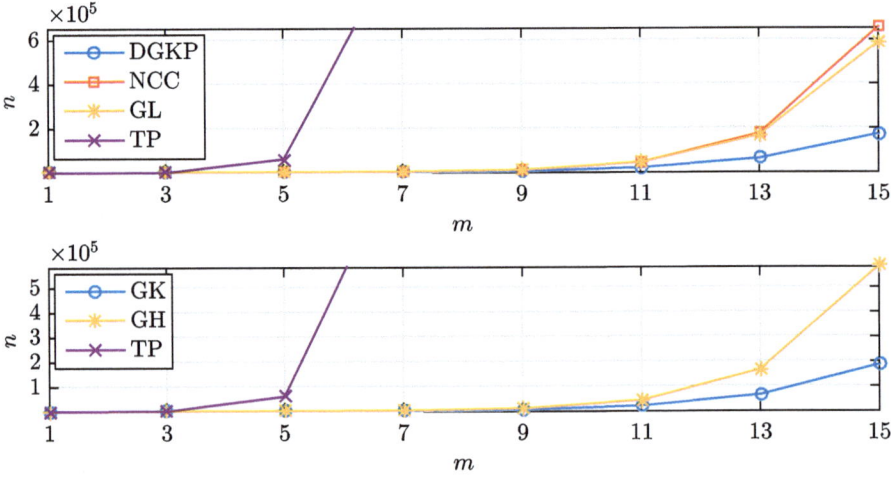

Fig. 3.3 Comparison of various underlying quadrature rules and tensor product for $d = 10$

3.2 Multidimensional Deterministic Numerical Integration

Again, the case of a Gaussian weight function is depicted in the lower and the case $w(x) = 1$ in the upper part of the figure. Here, the Smolyak cubature rules clearly use significantly less abscissae than the tensor product. For the case $w(x) = 1$ obviously the DGKP rules are most efficient. The NCC rules perform worst compared to the other rules. This is due to the fact that the cardinalities of the one-dimensional sets $S_1, S_2, \ldots$ must rise exponentially ($n_1 = 1$ and $n_i = 2^{(i-1)} + 1$) in order to provide a nested structure. So it seems that the positive effects of nestedness are compensated by the fast growth of the single sets. With respect to a certain range of dimensions and degrees of exactness a procedure to employ the nested Clenshaw–Curtis quadrature in a way that the resulting Smolyak cubature rules use significantly less abscissae has been found by Burkhardt (2014). In the case of the Gaussian weight function, as expected, the GH and the GK rules use much less abscissae than the tensor product. The best performing rules are the GK rules which use by far the lowest number of abscissae as d increases.

The main strength of the Smolyak cubature is illustrated in Fig. 3.4. Here the degree of exactness is set to $m = 13$ and the dimension varies from $d = 1$ to $d = 15$. For better visualization of the growth behaviour, the cardinalities of the sets of abscissae used by the Smolyak cubature rules and the tensor product, respectively, are plotted on logarithmic scale. While the number of abscissae used by the tensor product grows exponentially as d increases, the number of abscissae used by the Smolyak cubature rules always grows polynomially, no matter which underlying quadrature rule or weight function is chosen.

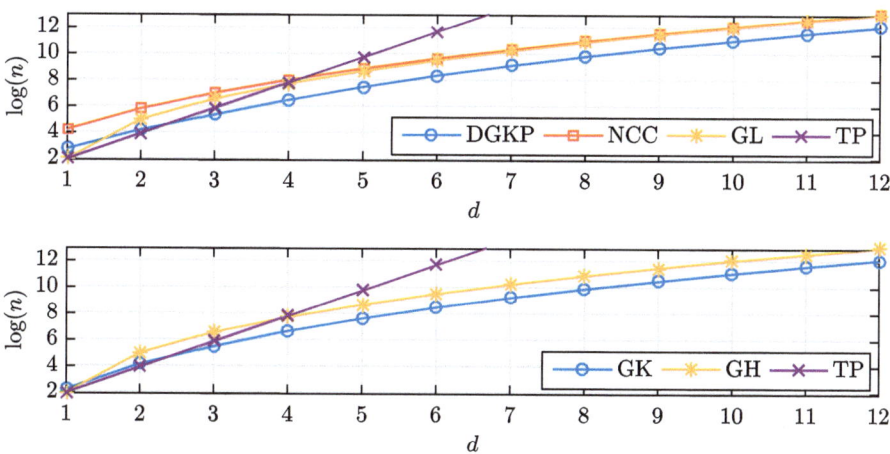

Fig. 3.4 Comparison of various underlying quadrature rules and tensor product for $m = 13$

3.2.7 Compound Rules

If the unweighted integral

$$\mathcal{I}[f] = \int_D f(x)\, dx \tag{3.155}$$

is so complicated that no sufficiently exact integration formula is available, compound rules represent an approach to increase the approximation accuracy (cf. Krommer and Ueberhuber 1998, p. 163). Given that the domain of integration is finite, it can often be decomposed into disjoint subregions $D_1, D_2, \ldots, D_k$, so that

$$\bigcup_{i=1}^{k} D_i = D. \tag{3.156}$$

The integral is then evaluated as

$$\mathcal{I}[f] = \sum_{i=1}^{k} \int_{D_i} f(x)\, dx. \tag{3.157}$$

The cubature rule of choice is therefore replicated k-times. Each replication then gets transformed for the application on one of the subregions by the method which will be described in Sect. 3.2.8. The approximation of the integral by the compound rule reads

$$\mathcal{I}[f] \approx \sum_{i=1}^{k} \sum_{l=1}^{n} f(\chi_{i,l})\, \alpha_{i,l}. \tag{3.158}$$

In the worst case, the compound rule uses $k \cdot n$ abscissae. This number can be reduced, if some of the abscissae are located on the boundary of a subregion (cf. Krommer and Ueberhuber 1998, p. 165). A special case of the compound rule, the so-called copy rule, emerges if all subregions are of the same shape (cf. Cools 1997, p. 2).

Figure 3.5 shows how the domain of integration gets decomposed into smaller subregions with regard to the copy rules which are used in this work. The accuracy of the rule can be influenced by changing the level parameter. Since the domain gets divided into 2^{level} subregions, the copy rule uses at most $2^{\text{level}} \cdot n$ abscissae.

3.2 Multidimensional Deterministic Numerical Integration

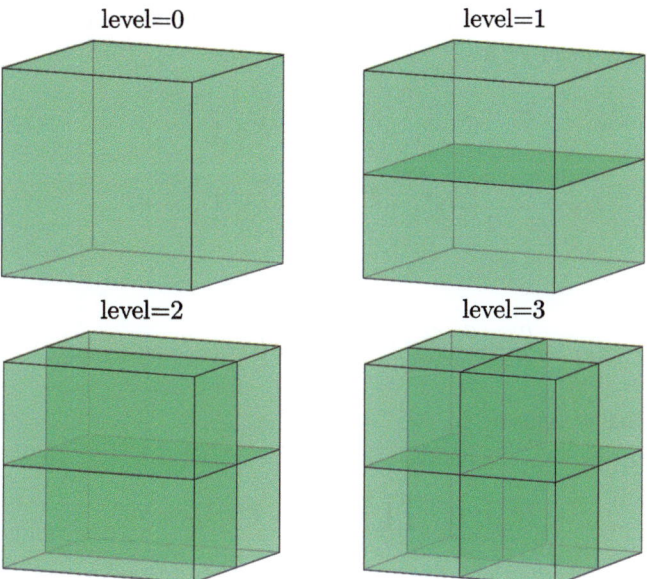

Fig. 3.5 Decomposition-levels

3.2.8 Change of Variables

A cubature rule for the domain $[-1, 1]^d$ can be used for integration on a different domain by a change of variables of the integral. Substituting the variables $y_1, y_2, \ldots, y_d$ of the integral

$$\int_{a_1}^{b_1} \int_{a_2}^{b_2} \cdots \int_{a_d}^{b_d} f(y_1, y_2, \ldots, y_d) \, dy \tag{3.159}$$

by

$$y_i = \frac{b_i - a_i}{2} x_i + \frac{a_i + b_i}{2}, \quad i = 1, \ldots, d \tag{3.160}$$

yields

$$\int_{a_1}^{b_1} \int_{a_2}^{b_2} \cdots \int_{a_d}^{b_d} f(y_1, y_2, \ldots, y_d) \, dy = \prod_{i=1}^{d} \frac{(b_i - a_i)}{2} \int_{-1}^{1} \int_{-1}^{1} \cdots \int_{-1}^{1} f(x_1, x_2, \ldots, x_d) \, dx. \tag{3.161}$$

The integral can therefore be approximated by the cubature rule after transforming the original abscissae χ_l^* to

$$\chi_{l,1} = \frac{b_1 - a_1}{2} \chi_{l,1}^* + \frac{a_1 + b_1}{2}, \ l = 1, \ldots, n$$

$$\chi_{l,2} = \frac{b_2 - a_2}{2} \chi_{l,2}^* + \frac{a_2 + b_2}{2}, \ l = 1, \ldots, n$$

$$\vdots$$

$$\chi_{l,d} = \frac{b_d - a_d}{2} \chi_{l,d}^* + \frac{a_d + b_d}{2}, \ l = 1, \ldots, n$$

(3.162)

and the original weights α_l^* to

$$\alpha_l = \alpha_l^* \cdot \prod_{i=1}^{d} \frac{(b_i - a_i)}{2}, \ l = 1, \ldots, n. \qquad (3.163)$$

The integral

$$\mathcal{I}[fw] = \int_{-\infty}^{\infty} f(y) \frac{1}{\sqrt{(2\pi)^d \det(\Sigma)}} e^{-\frac{1}{2}(y-\mu)'\Sigma^{-1}(y-\mu)} dy \qquad (3.164)$$

in which the weight function is the multivariate normal distribution with mean μ and covariance matrix Σ can be approximated using a cubature rule which is suitable in the case of the weight function

$$w(x) = \frac{1}{\sqrt{(2\pi)^d}} e^{-\frac{x'x}{2}}. \qquad (3.165)$$

This done by rewriting the integral in terms of uncorrelated random variables. Setting

$$y = \mu + \Gamma x, \qquad (3.166)$$

in which Γ is the lower triangular matrix of the Cholesky decomposition of Σ, yields

$$dy = \det(\Gamma) dx. \qquad (3.167)$$

3.2 Multidimensional Deterministic Numerical Integration

After the change of variables, the integral reads

$$\mathcal{I}[fw] = \int_{-\infty}^{\infty} f(\boldsymbol{\mu} + \boldsymbol{\Gamma}\boldsymbol{x}) \frac{1}{\sqrt{(2\pi)^d}} e^{-\frac{1}{2}\boldsymbol{x}'\boldsymbol{x}} d\boldsymbol{x}. \tag{3.168}$$

Hence, the integral can be approximated by the cubature rule after changing the original abscissae $\boldsymbol{\chi}_l^*$ to

$$\boldsymbol{\chi}_l = \boldsymbol{\mu} + \boldsymbol{\Gamma}\boldsymbol{\chi}_l^*, \; l = 1, \ldots, d. \tag{3.169}$$

Chapter 4
Optimization and Stabilization of Cubature Rules

In this chapter the methods used to generate own cubature rules are developed. The algorithm presented in the first chapter is based on a least squares approach to the multidimensional moment equations and is used to generate cubature rules with *approximate* exactness $m = 5$ and $m = 7$ which possess optimal stability. In the second section, three new kinds of cubature rules will be presented which are based on the Smolyak algorithm. The developed procedure allows the computation of Smolyak cubature rules of high stability.

4.1 Cubature Rules Based on a Least Squares Approach

To compute efficient and stable cubature rules for the case of centrally symmetric integrals, the simplified d-dimensional moment equations, $\boldsymbol{B}\boldsymbol{v} = \boldsymbol{c}$ (Sect. 3.2.5), can serve as a starting point. Instead of searching for abscissae which yield an exact solution to the system of equations, approximate solutions in the sense of least squares will be computed which meet specific quality criteria. Hence, the goal is to determine the fewest number of abscissae χ_l, for which the cubature rule has non-negative weights and sufficiently approximates a predetermined degree of exactness m. In addition it is required that the components of the abscissae, $a \leq \chi_{k,l} \leq b$, $k = 1, 2, \ldots, d$, $l = 1, 2, \ldots, n$, are located within certain limits. So strictly speaking, a cubature rule of this kind may be not exact for any polynomial. However, it is tried to keep the approximation error so small that it can be neglected in practical applications.

As initial value for n the number given by the lower bound of Möller (Sect. 3.2.2) will be used and the optimization will be carried out by using the MATLAB™ implementation of the interior-point algorithm[1] for nonlinear optimization which

---

[1] For an insight into interior-point methods, see Wright (2005).

is a gradient-based method. If the minimum of the objective function value is higher than a predefined target value, the parameter n will be increased and the optimization process starts again.

Several approaches are possible to find a solution to the given task, in which the restriction to strictly positive weights represents the greatest difficulty. A first idea is to treat the equation system

$$Bv = c \qquad (4.1)$$

as a non-negative least squares problem.[2] This leads to a nested optimization procedure and the objective function then is

$$\min_{\chi_{1,l},\chi_{2,l},\ldots,\chi_{d,l},\ l=1,2,\ldots,\lfloor n/2 \rfloor} \|Bv - c\|_2$$

subject to $a \leq \chi_{k,l} \leq b$, $k = 1, 2, \ldots, d$, $l = 1, 2, \ldots, \lfloor n/2 \rfloor$ \qquad (4.2)

and

$$\alpha_l \geq 0,\ l = 1, 2, \ldots, \lfloor n/2 \rfloor.[3]$$

Because within every iteration of the optimization procedure the vector of weights v is computed with respect to the constraint that all components of v are positive, the method is very time consuming and therefore not recommended.

Another way to suppress negative weights is applying a penalty function. The objective function to this approach reads

$$\min_{\chi_{1,l},\chi_{2,l},\ldots,\chi_{d,l},\ l=1,2,\ldots,\lfloor n/2 \rfloor} \|Bv - c\|_2 + \lambda \cdot \|v^-\|_1,\ \lambda > 0$$

subject to $a \leq \chi_{k,l} \leq b$, $k = 1, 2, \ldots, d$, $l = 1, 2, \ldots, \lfloor n/2 \rfloor$. \qquad (4.3)

In each iteration of the optimization process, given the current abscissae χ_l, the solution to the ordinary least squares problem, $\min \|Bv - c\|_2$, reads

$$v = (B'B)^{-1} B'c.[4] \qquad (4.4)$$

The vector v^- consists of *only* the negative entries of v and therefore the negative weights. By changing the parameter λ the effect of the penalty can be influenced which is a strength and a weakness at the same time. Often, finding the right level of λ which leads to strictly positive weights is very complicated. Moreover, the fixation

---

[2] For a very frequently used algorithm, see Lawson and Hanson (1995, p. 161).
[3] The notation $\lfloor x \rfloor$ is used in order to cover both of the cases of even and odd numbers of abscissae. The expression means that x is rounded to the largest integer not greater than x.
[4] This is the algebraic solution. However, in practical applications it is much more stable to use other methods, e.g. the QR-decomposition, in order to compute the solution.

4.1 Cubature Rules Based on a Least Squares Approach

of λ so that it can be used for all combinations of m and d does not seem to be possible. According to this, for each cubature rule a relatively large number of test runs are necessary in order to adjust λ in reasonable way. Another downside lies in the fact that the function values scale with λ and so the overall optimality criterion has to be adjusted every time λ is changed. Furthermore, the use of the penalty function, in dependence on λ, seems to change the topography of objective function in a negative way. The fact that many times the nonlinear optimizer gets stuck in a local minimum leads to the presumption that the penalty function increases the roughness of the objective function. In conclusion it can be stated that the penalty approach is indeed able to achieve good results, but at the cost of a greatly increased effort.

Third approach is based on enforcing positive weights indirectly. The objective function is formulated as

$$\min_{\chi_{1,l},\chi_{2,l},\ldots,\chi_{d,l},\ l=1,2,\ldots,\lfloor n/2\rfloor} \left\| \boldsymbol{B}\boldsymbol{v}^{+,0} - \boldsymbol{c} \right\|_2 \tag{4.5}$$

subject to $a \leq \chi_{k,l} \leq b$, $k = 1,2,\ldots,d$, $l = 1,2,\ldots,\lfloor n/2 \rfloor$.

The vector $\boldsymbol{v}^{+,0}$ denotes the vector $\boldsymbol{v}$, in which all negative components are set to zero. The implementation is straightforward. First, the solution to the ordinary least squares problem

$$\left\| \boldsymbol{B}\boldsymbol{v} - \boldsymbol{c} \right\|_2. \tag{4.6}$$

is calculated. Afterwards, the negative entries of $\boldsymbol{v}$ are set to zero which yields the vector $\boldsymbol{v}^{+,0}$. Using this new vector, the value of the objective function, $\left\| \boldsymbol{B}\boldsymbol{v}^{+,0} - \boldsymbol{c} \right\|_2$, is computed. In other words, within every iteration of the optimization procedure only the abscissae $\boldsymbol{\chi}_l$ whose corresponding weights are positive are used in the computation of (4.6). Therefore, in order to fully exploit the number of possible abscissae, the weights must be positive. For this reason, the quality of the approximation is indirectly connected to the number of positive weights. As neither a nested optimization procedure nor the introduction of additional parameters is necessary, this approach is the most efficient one of the three described methods. Furthermore, numerical experiments indicate that the use of this objective function leads to good results, with negative weights being reliably suppressed. For these reasons, this approach is used in the further course.

The target value must scale with the complexity of the approximation problem. In order to achieve this, as target value the number $\sqrt{p} \cdot 10^{-6}$ will be chosen. Here p is the dimension of the vector $\boldsymbol{c}$. Thus, the target value is equal to the norm of squared deviations, $\left\| \boldsymbol{B}\boldsymbol{v} - \boldsymbol{c} \right\|_2$, in the case that every squared deviation equals $(10^{-6})^2$. If the value of the objective function drops below $\sqrt{p} \cdot 10^{-6}$, the cubature rule is considered as sufficiently exact. If the minimum of the objective function is higher than the target value, the optimization will be restarted with an increased number of abscissae, n. The box restrictions, $a \leq \chi_{l,k} \leq b$, $l = 1, 2, \ldots, \lfloor n/2 \rfloor$, $k = $

$1, 2, \ldots, d$, will be set to $[-1, 1]$ for the unweighted case and to $[-5, 5]$ in the case of the Gaussian weight function.

Because of the high complexity of the emerging equation systems, the optimization approach in general is only applicable for a limited selection of degrees of exactness m and dimensions d. As an example, for the construction of a cubature rule of exactness $m = 5$ and dimension $d = 8$ the matrix B has 367 rows. Following the lower bound of Möller (Sect. 3.2.2), at minimum 73 abscissae are needed. Therefore B has at least $\lfloor n/2 \rfloor = 36$ columns. As mentioned, the midpoint is always part of the set of abscissae if n is odd. Hence, in this case one abscissa is fixed. So the objective function has to be minimized with respect to $\lfloor n/2 \rfloor \cdot d = 36 \cdot 8 = 288$ variables. For $m = 7$, B has 2,083 rows. The lower bound for the amount of abscissae in this case is 256. So the matrix B has $\lfloor n/2 \rfloor = 128$ columns and the objective function has to be minimized with respect to $\lfloor n/2 \rfloor \cdot d = 1,024$ variables. These examples show that the computational effort increases rapidly so that the optimization, at a certain point, is no longer possible. The effort is further increased by the fact that the required number of abscissae in many cases shows to be much higher than the number given by the lower bound. For these reasons, the optimization procedure will be applied to find cubature rules of a moderate degree of exactness, namely $m = 5$ and $m = 7$. For $m = 5$, cubature rules of dimensions $d = 2$ to $d = 12$ and for $m = 7$, cubature rules for dimensions $d = 2$ to $d = 8$ will be computed.

Tables 4.1 and 4.2 show the results of the optimization procedure for $w(x) = 1$ and the Gaussian weight function. The numbers of abscissae used by the cubature rules with an approximate degree of exactness which have been found by the optimization procedure are given in the column named AE. Next to this column, the most efficient alternative is given (Alt.). If the alternative cubature rule uses negative weights this is indicated by the letter N. In this case, also the numbers of abscissae used by most efficient alternative rule with strictly positive weights are shown (P-Alt.). Furthermore, the lower bound of Möller is given (LB).

For the unweighted case (Table 4.1), $m = 5$ and $d = 2$ to $d = 3$, the numbers of abscissae used by the best alternative rules equal the lower bound of Möller. Therefore, the AE rules yield no improvement in efficiency. For $m = 5$ and $d = 4$ to $d = 10$, the AE rules are more efficient than the alternative rules. The difference in used abscissae is especially striking for the dimensions $d = 8$ to $d = 10$. For the dimensions $d = 11$ and $d = 12$ it was not possible to generate a rule which is more efficient than the best alternative with negative weights. In comparison to the rules which use only positive weights, however, the AE rules are of superior efficiency. The AE rules for $m = 7$ use, except for $d = 2$, less abscissae than the alternative rules. For the dimensions $d = 5$ to $d = 8$ the AE rules have a further special feature, in the sense that there exists no alternative with strictly positive weights.

With respect to the Gaussian weight function (Table 4.2), the numbers of abscissae used by all cubature rules with $m = 5$ and for $d = 2$ to $d = 3$ equal the lower bound. Therefore, these rules possess optimal efficiency. The rules for $d = 4$ to $d = 7$ only use one more abscissa than the number given by the lower bound and the optimization approach does not yield any improvement. Good results have been obtained regarding the cubature rules for $d = 8$ to $d = 12$. For dimensions $d = 8$

4.1 Cubature Rules Based on a Least Squares Approach

Table 4.1 Cubature rules with approximate degree of exactness for the unweighted case

	$m=5$					
	n					
d	AE	Alt.		P-Alt.		LB
2	7	7		100		7
3	13	13		101		13
4	21	24		100		21
5	31	32		101		31
6	43	73		45		43
7	63	99		45		57
8	85	121	N	48	129	45	73
9	118	145	N	48	163	45	91
10	152	171	N	48	201	45	111
11	199	199	N	48	243	45	133
12	280	229	N	48	289	45	157
	$m=7$					
2	12	12		110		12
3	26	34		46		26
4	54	57		69		48
5	102	141	N	81	—	80
6	180	245	N	81	—	124
7	320	393	N	81	—	182
8	590	593	N	48	—	256

Table 4.2 Cubature rules with approximate degree of exactness in the case of the Gaussian weight function

	$m=5$					
	n					
d	AE	Alt.		P-Alt.		LB
2	7	7		103		7
3	13	13		104		13
4	22	22		101		21
5	32	32		101		31
6	44	44		101		43
7	57	57		101		57
8	90	121	N	48	144	114	73
9	118	145	N	48	146	114	91
10	156	171	N	48	225	114	111
11	195	199	N	48	278	114	133
12	245	229	N	48	280	114	157
	$m=7$					
2	12	12		104		12
3	27	27		104		26
4	49	49		102		48
5	82	83		102		80
6	129	137		102		124
7	208	227		102		182
8	551	673	N	48	768	102	256

to $d = 11$, the AE rules use less abscissae than the most efficient alternative rules although these rules were constructed without restricting the weights to positivity. Of particular note is the significant reduction of used abscissae in comparison to the best alternatives with strictly positive weights. For $d = 12$ it was not possible to find a cubature rule which is as efficient as the most efficient alternative. However, the computed rule uses significantly less abscissae than the best alternative with strictly positive weights. For $m = 7$ and $d \geq 5$, the AE rules show to be most efficient.

4.2 Construction of Stabilized Smolyak Cubature Rules

A weakness of the Smolyak cubature is that its stability decreases rapidly with increasing d. As the algorithm is based on combinations of tensor products of varying signs, negative weights arise even if the underlying quadrature rules have strictly positive weights.

The change of the stability factor SF (Sect. 3.2.1) for $m = 13$ and increasing d is shown in Fig. 4.1. As the dimension increases, the number of negative weights rises rapidly and the rules become more and more unstable. The stability factor of the product cubature rule was left out in the plot, because product cubature rules always have all-positive weights, if the underlying quadrature rules have positive weights. In this section, two methods for the stabilization of the Smolyak method will be elaborated and applied. The arising stabilized Smolyak cubature rules are referred to as stabilized(1) and stabilized(2).

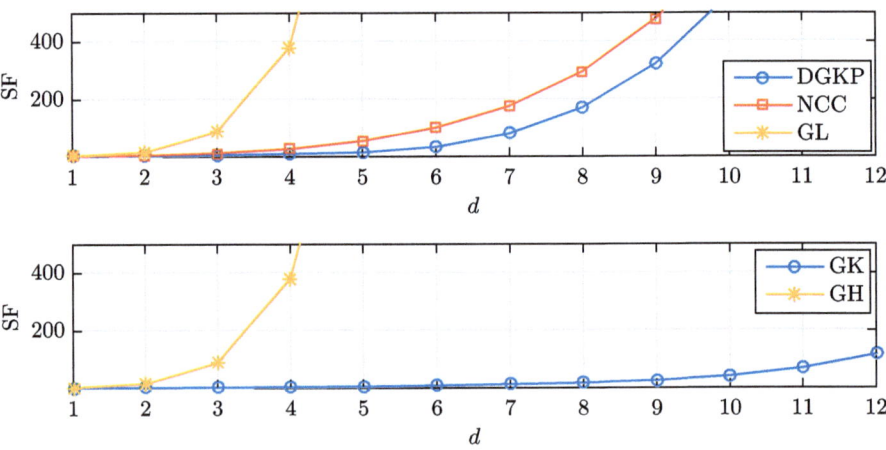

Fig. 4.1 Comparison of stability factors, $m = 13$

4.2.1 Stabilized(1) Rules

In order to construct own Smolyak cubature rules, first a set of quadrature rules has to be generated. This can be achieved by using the one-dimensional moment equations (3.12) in connection with the restrictions described in Sect. 3.1.2. A quadrature rule with odd degree of exactness m can be constructed by choosing $n = (m-1)/2$ abscissae. The weights ϑ_l are then uniquely determined by the arising system of equations.

As already been shown, the Smolyak algorithm delivers best results in terms of efficiency, if the cubature rules $Q_i, i = 1, 2, \ldots$, are nested. Therefore, it makes sense to define the sets of abscissae S_i, $i = 1, 2, \ldots$, used by the own quadrature rules, in the following way:

$$\begin{aligned}
S_1 &= \{0\} \ (m=1) \\
S_2 &= \{\chi_1, 0, -\chi_1\} \ (m=3) \\
S_3 &= \{\chi_1, \chi_2, 0, -\chi_2, -\chi_1\} \ (m=5) \\
S_4 &= \{\chi_1, \chi_2, \chi_3, 0, -\chi_3, -\chi_2, -\chi_1\} \ (m=7) \\
S_5 &= \{\chi_1, \chi_2, \chi_3, \chi_4, 0, -\chi_4, -\chi_3, -\chi_2, -\chi_1\} \ (m=9) \\
&\vdots
\end{aligned} \tag{4.7}$$

It is important to note that the sets are meant to be interlaced and therefore each set S_i contains all abscissae of its predecessor S_{i-1}. Furthermore, the requirement that $m_i \geq 2i - 1$ (Corollary in Sect 3.2.6) is fulfilled. Using the quadrature rules Q_i, $i = 1, \ldots, \frac{m+1}{2}$, which are based on the above sets of abscissae, as inputs for the Smolyak algorithm, yields a Smolyak cubature rule with new abscissae χ_l, weights α_l and exactness m.

The focus is now on the weights α_l and it is the goal to construct Smolyak cubature rules in which the influence of negative weights is as low as possible. The stability factor SF is used as the objective function to be minimized with respect to the variables $\chi_1, \chi_2, \ldots, \chi_n$:

$$\min_{\chi_1, \chi_2, \ldots, \chi_n} \frac{\sum_{l=1}^{n} |\alpha_l|}{\sum_{l=1}^{n} \alpha_l} \tag{4.8}$$

subject to $a \leq \chi_l \leq b$, $l = 1, 2, \ldots, n$.

After setting the values for m and d and the starting values for $\chi_1, \chi_2, \ldots, \chi_n$, every iteration of the optimization routine consists of three steps:

Step 1: Using the one-dimensional moment equations for the calculation of the weights ϑ_{l_i} of the quadrature rules Q_i, $i = 1, \ldots, (m-1)/2$, based on the sets S_i, $i = 1, 2, \ldots, (m-1)/2$.

Step 2: Construction of the Smolyak cubature by using Q_i, $i = 1, \ldots, (m-1)/2$, as inputs.

Step 3: Calculation of the stability factor SF.

For the unweighted case, the Smolyak cubature rules will be constructed for integrals over the domain $[-1, 1]^d$ and therefore the search region is restricted to $[-1, 1]$. In the case of the Gaussian weight function, again the interval $[-5, 5]$ will be chosen. Own numerical experiments have led to the assumption that the objective function is almost nowhere differentiable and has furthermore several local minima and maxima. Therefore, Newton's method and Quasi-Newton methods, respectively, cannot be used for optimization. Instead, a self-designed metaheuristic optimizer, called Random-Spheres-Optimizer, will be applied to find a satisfying solution to the optimization problem.

Metaheuristic optimization algorithms usually don't use the first or second derivative of the objective function. Instead, the search space is explored according to strategies which are often nature inspired and make use of stochastic elements (cf. Boussaïd et al. 2013, p. 82). Because of the complexity of the optimization problems, a global optimum can, in most cases, not be found. The aim is therefore rather to find a satisfactory solution within a predetermined period of time. Because optimization techniques are not in the focus of this work, the operation of the Random-Spheres-Optimizer will not be further discussed.[5]

Table 4.3 Comparison of stability factors in the case of the weight function $w(x) = 1$ and $m = 13$

d	DGKP	NCC	GL	Stabilized(1)
2	2.20	3.62	13.00	1.00
3	4.54	10.76	85.00	1.58
4	8.73	25.78	377.00	4.34
5	13.98	53.44	1,289.00	10.78
6	31.92	100.13	3,653.00	20.56
7	79.86	175.17	8,989.00	39.68
8	170.38	294.19	19,825.00	76.13
9	320.37	474.43	40,081.00	128.34
10	557.19	743.25	75,517.00	217.69
11	1,013.43	1,124.87	134,245.00	430.32
12	1,832.87	1,655.27	227,305.00	687.78
Δ %	⌀ 120.55%	⌀ 323.71%	⌀ 20,259.00%	

---

[5]The Random-Spheres-Optimizer can be downloaded at: http://www.mathworks.com/matlabcentral/fileexchange/48667-random-spheres-funct-startvalue-lb-ub-s-plotit-.

4.2 Construction of Stabilized Smolyak Cubature Rules

Table 4.3 shows the optimization results with respect to the weight function $w(x) = 1$. The obtained stability factors are compared to those of the DGKP, NCC as well as the GL rules. The degree of exactness has been set to $m = 13$ and the dimension varies from $d = 1$ to $d = 12$. Also the mean percentage deviations of the stability factors of the DGKP, NCC and GL rules from the stability factors of the own rules (stabilized(1)) are given and marked by $\Delta\%$. The stabilized(1) rules are characterized by being significantly more stable than the comparing Smolyak cubature rules. The difference is most prominently to be seen by the example of the stability factors connected to the GL rules. Here, the mean percentage deviation from the stability factors of the stabilized(1) rules is $20,259.00\%$. The DGKP rules produce the lowest stability factors of the three comparing rules but these still have an average deviation of 120.55% from the stability factors of the stabilized(1) rules.

The optimization results for the Smolyak cubature in the case of the Gaussian weight function are shown in Table 4.4. Furthermore, the stability factors connected to the GK rules and the GH rules are given and compared to those of the stabilized rule. Also in this table, the degree of exactness is $m = 13$ and d varies from $d = 1$ to $d = 12$. The stability factors obtained through optimization are even lower than in the case of the weight function $w(x) = 1$. Again, the stabilized(1) rules are of much higher stability than the comparing rules. In particular the difference in stability in comparison to the GH rule, which produces the same stability factors as that GL rules, is very noticeable.

Figure 4.2 shows the numbers of abscissae used by the comparison rules and the stabilized(1) rules. Again, the upper part of the figure shows cardinalities produced by the rules for the weight function $w(x) = 1$ while the lower part illustrates the cardinalities in the case of the Gaussian weight function. The degree of polynomial exactness is set to $m = 13$ and the dimension varies from $d = 1$ to $d = 12$. As far as the number of used abscissae is concerned, the stabilized(1) rules are not as efficient as the DGKP rules and the GK rules, respectively. This is due to the fact that the sets S_i (4.7) partly consist of more abscissae than the sets which are used as

Table 4.4 Comparison of stability factors in the case of the Gaussian weight function and $m = 13$

d	GK	GH	Stabilized(1)
2	1.13	13.00	1.00
3	1.92	85.00	1.00
4	2.63	377.00	1.00
5	4.03	1,289.00	1.04
6	7.20	3,653.00	1.57
7	11.83	8,989.00	2.34
8	16.59	19,825.00	3.61
9	24.19	40,081.00	5.14
10	39.00	75,517.00	8.12
11	66.58	134,245.00	14.08
12	114.69	227,305.00	24.75
$\Delta\%$	Ø 287.85%	Ø 447,089.00%	

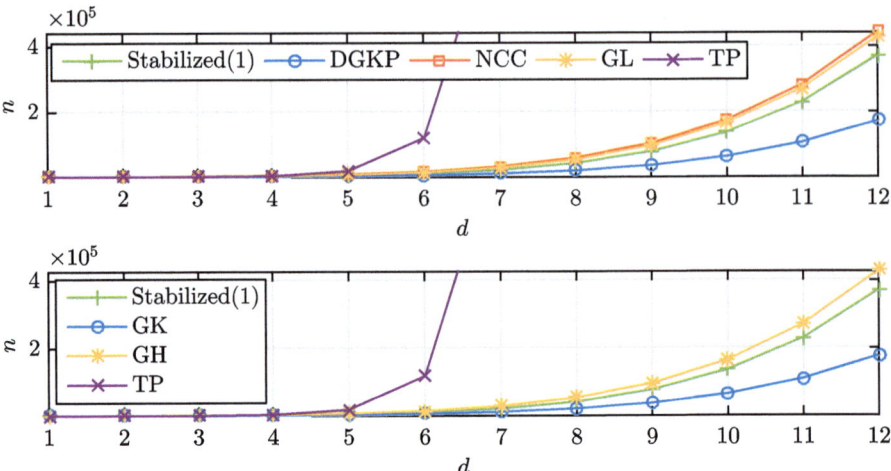

Fig. 4.2 Comparison of various underlying quadrature rules and tensor product to the stabilized rule for $m = 13$

a basis for the construction of the DGKP and GK rules. However, it should be noted positively that the stabilized(1) rules use less abscissae than the NCC, GL and GH rules.

4.2.2 Stabilized(2) Rules

A further leading approach to the stabilization of the Smolyak cubature is based on a slight change to the nested sets (4.7). This approach mainly delivers convincing results when applied to the Gaussian weight function. Therefore, the following discussion is intended to refer only to this case. As shown, using the Genz–Keister quadrature rules as input for the Smolyak algorithm leads to the most efficient Smolyak cubature rules. It would therefore be desirable to compute own Smolyak cubature rules of high stability while maintaining the efficiency of the GK rules.

In the first step the focus shall be on Smolyak cubature rules up to degree of $m = 9$. Heiss and Winschel (2008) use quadrature rules which use the following sets of abscissae and are based on the work of Genz and Keister (1996):

$$S_1 = \{0\} \; (m \geq 1)$$
$$S_2 = \{1.73, 0, -1.73\} \; (m \geq 3)$$
$$S_3 = \{1.73, 0, -1.73\} \; (m = 5) \tag{4.9}$$
$$S_4 = \{0.74, 4.18, 1.73, 0, -1.73, -4.18, -0.74\} \; (m = 7)$$
$$S_5 = \{2.86, 0.74, 4.18, 1.73, 0, -1.73, -4.18, -0.74, -2.86\} \; (m \geq 9)$$

4.2 Construction of Stabilized Smolyak Cubature Rules

These sets represent the most efficient solution to the requirement that the quadrature rules Q_i must have the degrees of exactness $m_1 \geq 2i - 1$ and be nested. Taking a look at the cardinalities,

$$
\begin{aligned}
\#S_1 &= 1 \\
\#S_2 &= 3 \\
\#S_3 &= 3 \\
\#S_4 &= 7 \\
\#S_5 &= 9.
\end{aligned}
\quad (4.10)
$$

it is obvious that the set S_3 of (4.7) contains two more abscissae in comparison to the set S_3 of (4.10). This results in a higher number of cubature abscissae with respect to the self-developed rules. By introducing the first three Gauss–Hermite abscissae into (4.7), the structure changes to

$$
\begin{aligned}
S_1 &= \{0\} \ (m = 1) \\
S_2 &= \{1.73, 0, -1.73\} \ (m \geq 3) \\
S_3 &= \{1.73, 0, -1.73\} \ (m = 5) \\
S_4 &= \{\chi_1, \chi_2, 1.73, 0, -1.73, -\chi_2, -\chi_1\} \ (m = 7) \\
S_5 &= \{\chi_1, \chi_2, \chi_3, 1.73, 0, -1.73, -\chi_3, -\chi_2, -\chi_1\} \ (m = 9)
\end{aligned}
\quad (4.11)
$$

and, up to degree $m = 9$, the cardinalities of the sets coincide with those used by Heiss and Winschel (2008). For the construction of rules with an exactness of $m = 7$ and $m = 9$ there are two and three, respectively, variables available which can be used to enhance the stability of the cubature rule. The number of resulting abscissae in both cases *equals* the number used by the GK rules. Because the use of two variables which are available for the case $m = 7$ only leads to a slight stabilization, the approach will be demonstrated for $m = 9$.

In Table 4.5 the two stabilization methods are compared to each other. The first method (stabilized(1)) yields a much stronger increase in stability than method two (stabilized(2)). This is due to the fact that the first stabilization approach offers more variables which, however, also leads to a significant increase of abscissae. Although the stabilized(2) rules are not as stable as the stabilized(1) rules, with a mean percentage deviation from the stability factors produced by the GK rules of over 170%, the improvement is still very clear. The main advantage of the stabilized(2) rules lies in the fact that the numbers of abscissae used are as low as those used by the GK rules.

The second stabilization method also has positive effects when constructing Smolyak cubature rules of exactness $m > 9$. This will be shown by the example of Smolyak cubature rules of degree $m = 11$. The six sets used by Heiss and Winschel

Table 4.5 Comparison of stabilization methods, $m = 9$

Dim	GK		Stabilization (1)		Stabilization(2)	
	n	SF	n	SF	n	SF
2	37	1.48	41	1.00	37	1.00
3	93	2.64	129	1.00	93	1.00
4	201	3.84	321	1.00	201	1.05
5	401	4.95	681	1.01	401	1.22
6	749	6.07	1,289	1.26	749	1.26
7	1,317	8.50	2,241	1.56	1,317	2.43
8	2,193	14.52	2,649	1.96	2,193	5.46
9	3,481	23.82	5,641	2.32	3,481	11.14
10	5,301	37.28	8,361	2.83	5,301	20.51
11	7,789	56.18	11,969	3.41	7,789	34.77
12	11,097	82.00	16,641	4.20	11,097	55.45
Δ% (1)	Ø717.95%					
Δ% (2)	Ø171.45%					

(2008) to compute GK rules of degree $m = 11$ read

$$S_1 = \{0\} \ (m = 1)$$
$$S_2 = \{1.73, 0, -1.73\} \ (m \geq 3)$$
$$S_3 = \{1.73, 0, -1.73\} \ (m = 5)$$
$$S_4 = \{0.74, 4.18, 1.73, 0, -1.73, -4.18, -0.74\} \ (m \geq 7) \quad (4.12)$$
$$S_5 = \{2.86, 0.74, 4.18, 1.73, 0, -1.73, -4.18, -0.74, -2.86\} \ (m \geq 9)$$
$$S_6 = \{2.86, 0.74, 4.18, 1.73, 0, -1.73, -4.18, -0.74, -2.86\} \ (m \geq 11)$$

and have the cardinalities

$$\#S_1 = 1$$
$$\#S_2 = 3$$
$$\#S_3 = 3$$
$$\#S_4 = 7 \quad (4.13)$$
$$\#S_5 = 9$$
$$\#S_6 = 9.$$

4.2 Construction of Stabilized Smolyak Cubature Rules

The set which is used in the second optimization approach reads

$$\begin{aligned}
S_1 &= \{0\} \ (m = 1) \\
S_2 &= \{1.73, 0, -1.73\} \ (m = 3) \\
S_3 &= \{1.73, 0, -1.73\} \ (m = 5) \\
S_4 &= \{\chi_1, \chi_2, 1.73, 0, -1.73, -\chi_2, -\chi_1\} \ (m = 7) \\
S_5 &= \{\chi_1, \chi_2, \chi_3, 1.73, 0, -1.73, -\chi_3, -\chi_2, -\chi_1\} \ (m = 9) \\
S_6 &= \{\chi_1, \chi_2, \chi_3, \chi_4, 1.73, 0, -1.73, -\chi_4, -\chi_3, -\chi_2, -\chi_1\} \ (m = 11).
\end{aligned} \quad (4.14)$$

As can be seen, the set S_6 of (4.14) uses two more abscissae than the set S_6 of (4.12). Thus, for $m = 11$, the stabilized(2) rules will use more abscissae than the GK rules. Fortunately, the efficiency of the optimized rules is hardly adversely affected by this disadvantage.

The optimization results are shown in Table 4.6. As was to be expected, the first optimization approach yields by far not only the most stable Smolyak cubature rules but also the most inefficient ones. The stabilized(2) rules, however, exhibit a very favourable ratio between efficiency and stability. The stability factors of the GK rules on the average are over 155% higher than those of the stabilized(2) rules. The difference in terms of used abscissae, however, is negligible.

Table 4.6 Comparison of stabilization methods, $m = 11$

d	GK		Stabilized(1)		Stabilized (2)	
	n	SF	n	SF	n	SF
2	45	1.26	61	1.00	49	1.00
3	165	2.38	231	1.00	171	1.02
4	441	3.53	681	1.00	449	1.37
5	993	5.02	1,683	1.03	1,003	1.80
6	2,021	7.53	3,653	1.28	2,033	1.97
7	3,837	9.70	7,183	1.99	3,851	2.14
8	6,897	13.37	13,073	2.80	6,914	3.20
9	11,833	23.30	22,363	3.89	11,851	6.35
10	19,485	40.38	36,365	4.83	19,505	13.35
11	30,933	68.46	56,695	6.33	30,955	26.33
12	47,529	111.05	85,305	12.61	47,553	48.77
Δ % (1)	∅459.52%					
Δ % (2)	∅155.72%					

4.2.3 Smolyak Cubature Rules with an Approximate Degree of Exactness

The free parameters which are used in both stabilization approaches can also be used to additionally decrease the number of used abscissae. The steps of the algorithm given in Sect. 4.2.1 which is used for the construction of the stabilized(1) and stabilized(2) rules remain mainly unchanged. Only the objective function is different. The idea here is to shrink a fraction τ, $0 \leq \tau \leq 1$, of the weights towards zero while maintaining a higher level of stability than the corresponding GK rule. A successful optimization is characterized by the achievement of two objectives:

1. After sorting the n absolute values of the weights in ascending order, the sum of the first c values must be so small in relation to the sum of all absolute weights that these weights are negligible.
2. The stability factor SF of the new rule must be smaller than the stability factor of the GK rule of the same dimension d and with the same exactness m, denoted as SF_{GK}.

This will now be presented by using the sets of the second stabilization approach (4.14) and the Gaussian weight function. As degrees of exactness, $m = 9$ and $m = 11$ will be chosen. The objective function reads

$$\min_{\chi_1, \chi_2, \dots, \chi_n} \frac{\sum_{l=1}^{c} \alpha_l^s}{\sum_{l=1}^{n} |\alpha_l|}, \quad c = \lfloor n \cdot \tau \rfloor, \; 0 \leq \tau \leq 1$$

subject to $a \leq \chi_l \leq b$, $l = 1, 2, \dots, n$ (4.15)

and

$$\sum_{l=c+1}^{n} \alpha_l^s < SF_{GK} (1 - \lambda), \; 0 \leq \lambda \leq 1.$$

The α_l^s, $l = 1, 2, \dots, n$, represent the absolute values of the weights, sorted in ascending order. To control the minimum level of stability, the parameter λ is introduced. More specifically, the stability factor of the optimized rule is not allowed to be greater than a certain percentage $1 - \lambda$ of the stability factor of the corresponding GK rule (SF_{GK}). The algorithmic implementation of this lower bound is very simple. In the cases where $SF \geq SF_{GK} (1 - \lambda)$, the value 10^{16} is assigned to the objective function. By doing this, the optimization algorithm is forced to search regions for which $SF < SF_{GK} (1 - \lambda)$. Because gradient-based optimization routines don't lead to satisfactory results when applied to the given objective function, the optimization will again be carried out by using the Random-Spheres-Optimizer (see Sect. 4.2.1). Generally, many different combinations of τ and λ are possible and worth to be tested, but for reasons of simplicity, the parameter

4.2 Construction of Stabilized Smolyak Cubature Rules

λ will be specified as $\lambda = \tau$. As range for the abscissae χ_l, again $[a,b] = [-5,5]$ will be chosen.

The downside of the optimization procedure is that it is not possible to fully control the numerical exactness of the cubature rule which is to be optimized. Even if some of the weights are extremely close to zero, the omission of these weights will lead to a violation of the moment equations and as a consequence, the resulting Smolyak cubature rule will only have an degree of exactness of *approximately m*. In previous test it has shown that the objective function must take on extremely small values ($< 10^{-10}$) to ensure that the resulting cubature rule is still exact enough. Thus, the optimization stops if the objective function takes on a value smaller than 10^{-10}. If no solution is obtainable, the parameter τ gets adjusted and the optimization restarts.

Table 4.7 shows the optimization results for $d = 2$ to $d = 10$ and the degrees of exactness $m = 9$ and $m = 11$. The optimized Smolyak cubature rules, as explained, only have an approximate degree of exactness (AE) and are therefore called Smolyak-AE rules. In the second column, the abscissae used by the Smolyak-AE and the GK rules are given. Furthermore, the percentage differences of the numbers of used abscissae are shown in the third column. The stability factors of the Smolyak-AE and the GK rules are given in the fourth column. Column five shows the percentage difference of the stability factors. In order to determine the error

Table 4.7 Comparison of Smolyak-AE and Smolyak cubature rules, $m = 9$ and $m = 11$

d	Smolyak-AE (n)	GK (n)	Δ %	SF Smolyak-AE	SF GK	Δ %	Error norm
m = 9							
2	32	37	15.62%	1	1.48	48.00%	1.9346e-09
3	80	93	16.25%	2.23	2.64	18.39%	2.5206e-07
4	177	201	13.56%	3.38	3.84	13.61%	1.8831e-10
5	361	401	11.08%	1.49	4.95	232.21%	1.2815e-09
6	532	749	40.79%	4.28	6.07	41.12%	8.1611e-10
7	1,238	1,317	6.38%	7.28	8.50	16.76%	4.8125e-09
8	2,084	2,193	5.23%	13.16	14.52	10.33%	–
9	3,351	3,481	3.88%	20.79	23.82	14.57%	–
10	5,142	5,301	3.09%	35.79	37.28	4.16%	–
m = 11							
2	40	45	12.50%	1	1.26	26.00%	2.2890e-05
3	141	165	17.02%	1.79	2.38	32.96%	1.1468e-06
4	396	441	11.36%	2.81	3.53	25.62%	2.8055e-05
5	853	993	17.79%	4.24	5.02	18.40%	–
6	1,891	2,021	6.87%	7.00	7.53	7.57%	–
7	3,043	3,837	26.09%	7.66	9.70	26.63%	–
8	6,499	6,897	6.12%	10.16	13.37	31.59%	–
9	11,259	11,833	5.10%	21.49	23.30	8.42%	–
10	18,530	19,485	5.15%	39.97	40.38	1.03%	–

of the Smolyak-AE rules with respect to the violation of the moment equations, the norm of the error vectors have been calculated and are given in the outer right column. Due to limited computing capacity, this calculation was only possible for specific dimensions and degrees of exactness.

The Smolyak-AE rules are all of higher efficiency and stability than the GK rules. The highest combined improvement has been obtained for $d = 6$ and $m = 9$ where the GK rule uses over 40% more abscissae than the Smolyak-AE rule. Also the stability factor is over 40% higher than the one of the corresponding Smolyak-AE rules. As can be seen, the error of the Smolyak-AE rules is very small in relation to the number of moment equations (5005 for $d = 6$ and $m = 9$) and can therefore very probably be neglected in practical applications.

For $m = 9$ and $d \leq 4$ as well as $m = 11$ and $d \leq 5$ the most efficient cubature rules have been proposed by Stenger (cf. Stenger 1971, pp. 7–9). Considering higher dimensions, no cubature rules are known which are more efficient than the GK rules. The Smolyak-AE rules outperform all mentioned rules in terms of the number of used abscissae. Thus, treating the Smolyak-AE rules as exact, these represent the most efficient rules available for the degrees of exactness $m = 9$ and $m = 11$ for dimensions $d = 2$ to $d = 10$.

Chapter 5
Simulation Studies

The efficiency and stability of the AE and Smolyak-AE cubature rules in comparison to the GK and DGKP rules will be investigated by conducting four simulation studies. In the first two studies, five- and seven-dimensional state-space models will be filtered by the nonlinear Kalman filter, which operates analogously to the unscented Kalman filter (Algorithm 3). In the third study, the conditional Kalman filter (Algorithm 5) will be applied. The algorithms of the unscented Kalman filter and the conditional unscented Kalman filter can be easily equipped with arbitrary cubature rules which are suitable for Gaussian integrals. In order to achieve this, only the sections "Definitions" and "Initialization" have to be changed using the change of variables approach for Gaussian integrals described in Sect. 3.2.8. The modifications are here presented by the example of the unscented Kalman filter algorithm. To make clear that the algorithm is used in connection with different types of cubature rules, its name is changed to "cubature-based Kalman filter algorithm" in the following example.

Algorithm 7 The cubature-based Kalman filter algorithm
1: **procedure** CBKF
2:     *Definitions*:
3:     $\Gamma = \text{chol}(\Sigma)$, lower
4:     $\chi(\mu, \Sigma) = \mu + \Gamma \chi_l^*$, $l = 1, \ldots, n$
5: _____
6:     *Initialization*:
7:     $\mu_{y,1|0}$; $\Sigma_{yy,1|0}$; $\chi^* = [\chi_1^*, \ldots, \chi_n^*]$; $\alpha = [\alpha_1, \ldots, \alpha_n]$
8: _____
9:     $\vdots$
10: **end procedure**

© Springer International Publishing AG 2017
D. Ballreich, *Stable and Efficient Cubature-based Filtering in Dynamical Systems*, DOI 10.1007/978-3-319-62130-2_5

Also the discrete Bayes filter (Algorithm 1), which will be used in the fourth simulation study, can be equipped with arbitrary cubature rules apart from Riemann sums. These rules must be suitable for unweighted integrals. As already mentioned, the integration limits must be set before the start of the algorithm. Therefore the cubature rule of choice, which will generally be given for the integration on the domain $[-1, 1]^d$, has to be transformed by using the formulas (3.162) and (3.163) beforehand.

Within the nonlinear Kalman filter various approximations influence each other. First, the nonlinear Kalman filter is "only" the best *linear* approximation to the Bayes filter (Sect. 2.3.2). However, it must be noted that this only applies if the densities $p\left(\mathbf{y}_t|\mathbf{Z}^{t-1}\right)$ and $p\left(\mathbf{y}_t|\mathbf{Z}^t\right)$ are known, which is not the case in practice. Therefore, the next approximation is due to the fact that these densities are replaced by Gaussians (Sect. 2.3.2). Finally, at least in the higher-dimensional case, for practical applications only approximations to the values of the Gaussian integrals are efficient in terms of computation time. This multitude of different types of approximations makes it hard to deduce statements regarding the quality of a certain cubature rule. In fact, if the real densities $p\left(\mathbf{y}_t|\mathbf{Z}^{t-1}\right)$ and $p\left(\mathbf{y}_t|\mathbf{Z}^t\right)$ deviate strongly from their Gaussian counterparts, which is usually the case in highly nonlinear models, increasing the exactness of the cubature rule does not necessarily lead to a better approximation to the state. Quite the contrary occurs, as even a cubature rule of low accuracy can, on the average, provide a better approximation to the state than a rule of high exactness within a simulation study due to circumstances that have nothing to do with the accuracy of the cubature rule. As an example, a certain parametrization of the drift function or the length of the simulated time series can be mentioned.

In order to still being able to make meaningful comparisons, the procedure which will be applied in the first two simulation studies is as follows: First, each simulation study will be carried out using a cubature rule with very high exactness in order to approximate the converged filter solutions. These solutions are characterized by the fact that all the required Gaussian integrals have been calculated almost exactly within every iteration of the filter algorithm. Increasing the exactness of the used cubature rule even further therefore would not lead to a significant change of the results. On the basis of these filter solutions, the mean absolute deviation (MAD) to the simulated states will be calculated. This converged MAD (MAD_C) is then compared to the MADs generated by the cubature-based nonlinear Kalman filter, equipped with cubature rules of lower exactness. The absolute percentage difference between the MAD_C and the MADs resulting from the use of the AE, Smolyak-AE and GK rules can be interpreted as indirect performance measure. The aim is to investigate whether the AE and Smolyak-AE rules lead to an as fast convergence to the MAD_C as the GK rules, while being more efficient. As a standard tool, also the cubature Kalman filter (CKF, UKF with $\kappa = 0$) will be included in the comparison.

As already mentioned, all parameters used in a state-space model exert a potentially large influence on the outcome of a simulation study. Consequently, to be able to generalize about the results more than it would be possible by using a fixed parametrization, it would be reasonable to perform multiple simulation studies with various different parametrizations and to average the results. Due to the vast number of possible combinations of parameters, this is hardly feasible. With respect to the state-space models used within the first two simulation studies, the diffusion function is not dependent on y_t. As a consequence no integration of Ω is necessary, and thus this part of the state-space model is not directly influenced by the chosen cubature rule. Therefore, and in order to keep the computational complexity at a manageable level, the variation of different parameters will be limited to the variation of the parameters of the drift function. The parameter values will be chosen randomly from predetermined intervals in each simulation. Accordingly, this variability makes it necessary to generate a much higher number of replications as it normally would be sufficient. Therefore, the number of replications is set to 500,000.

The third simulation study, in which the conditional Kalman filter (Algorithm 5) is applied to a nine-dimensional state-space model is the computationally most demanding one. Again, the framework of the presented algorithm stays unchanged and only various sets of abscissae and weights are used and examined for differences in performance. Because of the high dimensionality of the state space, the conduction of a detailed simulation study in combination with the variation of the parameters of the drift function and the calculation of the converged filter solutions is not feasible within a reasonable time. Thus, only one parametrization will be used and the solutions produced by the different cubature rules will be compared. It must be stressed that with this study setup no profound statements regarding the general quality of the cubature rules can be made, but it is possible to compare, if the AE and Smolyak-AE rules deliver similar results, measured as MAD, as the GK while using less computational resources. As in the first two simulation studies, additionally to the AE, Smolyak-AE and GK rules, also the CKF will be applied. The number of simulation runs is set to 1,000.

In the fourth simulation study the discrete Bayes filter (Algorithm 1) will be used to filter a three-dimensional state-space model in order to compare the filter performance of the Riemann abscissae and weights to the performance of the AE and DGKP rules. Because the discrete Bayes filter is based on the true densities $p\left(y_t|\mathbf{Z}^{t-1}\right)$ and $p\left(y_t|\mathbf{Z}^t\right)$ here a comparison of the MADs as measure of the filtering accuracy makes sense and the detour via the MAD_C is unnecessary. Like in simulation studies one and two, the parameters of the drift function will be chosen randomly from a predefined interval in each simulation run and the number of replications is set to 15,000.

5.1 The Univariate Non-Stationary Growth Model

The univariate non-stationary growth model has gained wide popularity as a test model for nonlinear filters and has been, inter alia, examined by Kitagawa (1987) and Carlin et al. (1992):

$$y_t = \phi_1 y_{t-1} + \phi_2 \frac{y_{t-1}}{1 + y_{t-1}^2} + \phi_3 \cos(1.2(t-1)) + g\epsilon_t$$

$$z_t = \frac{1}{20} y_t^2 + \delta_t.$$
(5.1)

Inspecting only the iterative part of the drift function,

$$y_t = \phi_1 y_{t-1} + \phi_2 \frac{y_{t-1}}{1 + y_{t-1}^2},$$
(5.2)

provides information about the dynamics of the system which can be visualized by a cobweb plot (Fig. 5.1). As two exemplary starting points for the iteration, -0.2 and 2 have been chosen and the parameter values read $\phi_1 = 0.5$ and $\phi_2 = 25$. Starting from these points (triangles) and moving on with the iteration, the function values are attracted by two fixed points, marked by squares.

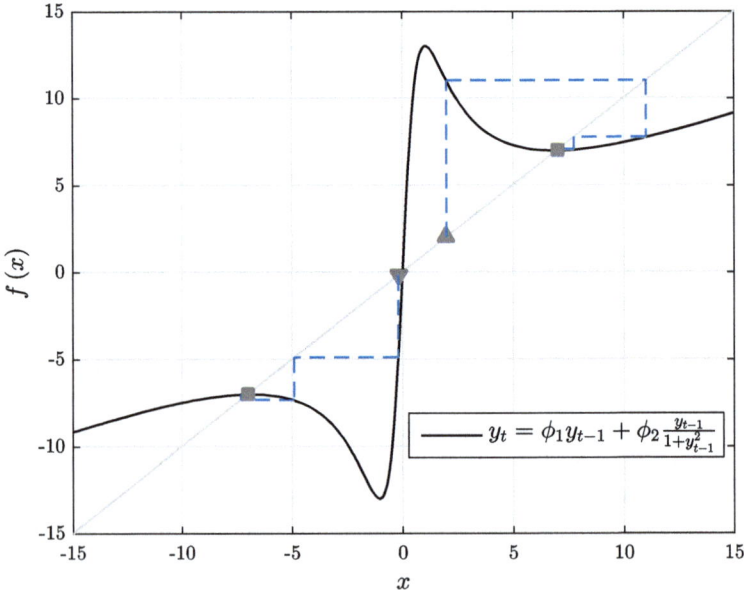

Fig. 5.1 Cobweb plot

5.1 The Univariate Non-Stationary Growth Model

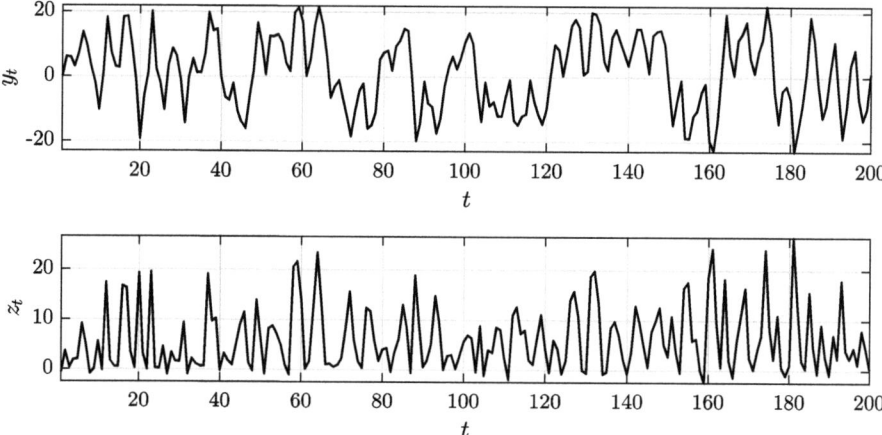

Fig. 5.2 The univariate non-stationary growth model

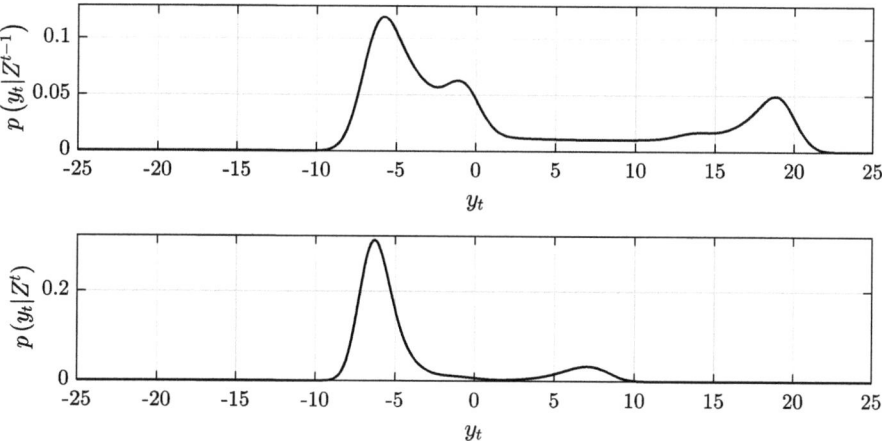

Fig. 5.3 Exemplary prior and posterior density of the non-stationary growth model

Figure 5.2 shows a simulated non-stationary growth model with $\phi_1 = 0.5$, $\phi_2 = 25$, $\phi_3 = 8, g = \sqrt{10}$ and $\mathbb{V}[\delta_t] = 1$. The strong influence of (5.2) is displayed by the fact that the values of the system model change constantly between two regimes. Furthermore it is recognizable that the measurements give only very little information about the position of the state.

As a consequence of the fact that only the square of y_t is measurable, the measurement density is bimodal in every time step. The posterior density is directly and the prior density is indirectly influenced by the measurement density (see Sect. 2.2). As a result, bimodal prior and posterior densities occur. These are exemplary shown in Fig. 5.3. This makes the model very hard to be filtered by a nonlinear Gaussian filter.

In order to extend the non-stationary growth model into five dimensions, the model will be replicated in each dimension and then coupled by correlated errors. Thus, the state-space model used for the simulation study reads

$$\begin{bmatrix} y_{1,t} \\ y_{2,t} \\ y_{3,t} \\ y_{4,t} \\ y_{5,t} \end{bmatrix} = \begin{bmatrix} \phi_1 y_{1,t-1} + \phi_2 \frac{y_{1,t-1}}{1+y_{1,t-1}^2} + \phi_3 \cos(1.2(t-1)) \\ \phi_1 y_{2,t-1} + \phi_2 \frac{y_{2,t-1}}{1+y_{2,t-1}^2} + \phi_3 \cos(1.2(t-1)) \\ \phi_1 y_{3,t-1} + \phi_2 \frac{y_{3,t-1}}{1+y_{3,t-1}^2} + \phi_3 \cos(1.2(t-1)) \\ \phi_1 y_{4,t-1} + \phi_2 \frac{y_{4,t-1}}{1+y_{4,t-1}^2} + \phi_3 \cos(1.2(t-1)) \\ \phi_1 y_{5,t-1} + \phi_2 \frac{y_{5,t-1}}{1+y_{5,t-1}^2} + \phi_3 \cos(1.2(t-1)) \end{bmatrix} + \boldsymbol{g}\epsilon_t$$

$$z_t = \frac{1}{20} \begin{bmatrix} y_{1,t}^2 \\ y_{2,t}^2 \\ y_{3,t}^2 \\ y_{4,t}^2 \\ y_{5,t}^2 \end{bmatrix} + \boldsymbol{\delta}_t. \tag{5.3}$$

The covariance matrices $\boldsymbol{\Omega}$ and $\boldsymbol{R}$ of the system errors and the measurement errors will be set constant and read

$$\boldsymbol{\Omega} = \boldsymbol{gg}' = \begin{bmatrix} 10 & 1 & 1 & 1 & 1 \\ 1 & 10 & 1 & 1 & 1 \\ 1 & 1 & 10 & 1 & 1 \\ 1 & 1 & 1 & 10 & 1 \\ 1 & 1 & 1 & 1 & 10 \end{bmatrix} \tag{5.4}$$

and

$$\boldsymbol{R} = \begin{bmatrix} 1 & 0 & 0 & 0 & 0 \\ 0 & 1 & 0 & 0 & 0 \\ 0 & 0 & 1 & 0 & 0 \\ 0 & 0 & 0 & 1 & 0 \\ 0 & 0 & 0 & 0 & 1 \end{bmatrix}. \tag{5.5}$$

The state-space model will be filtered with cubature rules of exactness $m = 5$, $m = 7$, $m = 9$ and $m = 11$. For each case, the results generated by the most efficient and stable self-developed rules afterwards are compared with the results delivered by the GK rule of the same exactness. For $m = 5$ and $m = 7$ the most efficient and stable own rules are the AE rules (Table 4.2) and for $m = 9$ and $m = 11$ the Smolyak-AE rules (Table 4.7). Additionally, as a standard filter algorithm, the CKF ($m = 3$) will be applied. To achieve an appropriate accuracy in the calculation of the MAD$_C$, a stabilized Smolyak cubature rule of exactness $m = 17$ (stabilized (1)),

5.1 The Univariate Non-Stationary Growth Model

Sect. 4.2.1) will be used which has a very high stability of SF $= 1.08$ and uses $n = 13{,}073$ abscissae.

$K = 500{,}000$ time series of length $T = 100$ will be generated with the initial state $\boldsymbol{y}_1 = \begin{bmatrix} 0 & 0 & 0 & 0 & 0 \end{bmatrix}'$. The initial prior state covariance matrix $\boldsymbol{\Sigma}_{yy,1|0}$ is set to

$$\boldsymbol{\Sigma}_{yy,1|0} = \begin{bmatrix} 1000 & 0 & 0 & 0 & 0 \\ 0 & 1000 & 0 & 0 & 0 \\ 0 & 0 & 1000 & 0 & 0 \\ 0 & 0 & 0 & 1000 & 0 \\ 0 & 0 & 0 & 0 & 1000 \end{bmatrix} \qquad (5.6)$$

and the initial prior state estimate $\boldsymbol{\mu}_{1|0}$ will be randomly drawn from $\mathcal{N}\left(\boldsymbol{y}_1, \boldsymbol{\Sigma}_{yy,1|0}\right)$. Furthermore, the parameters ϕ_1, ϕ_2 and ϕ_3 will be chosen from the uniform distributions $U(-0.9, 0.9)$, $U(-50, 50)$ and $U(-10, 10)$. The performance measures to be calculated are the MAD,

$$\mathrm{MAD}^1 = \frac{1}{K} \sum_{k=1}^{K} \frac{\frac{1}{T} \sum_{j=1}^{5} \left\| y_j^k - \mu_j^k \right\|_1}{5}, \qquad (5.7)$$

and the average time per filter run in seconds.

Table 5.1 shows the results of the simulation study in terms of the MADs. First, it is striking that, due to the strong nonlinearity of the state-space model, the increase of the exactness of the used cubature rules does not lead to improved filter performances. In fact, the opposite is the case. Starting from a MAD of 7.07 produced by the CKF, the MAD drops down to a lower level and from then on increases with higher degree of exactness m to the convergence value MAD_C. This applies to both groups of cubature rules, AE/Smolyak-AE and GK rules and is a clear indication for strongly non-Gaussian filter densities. To compare the performances to each other, the absolute percentage deviations of the MADs to the MAD_C, which are shown in the columns "$\Delta\%$", serve as indirect quality measures. Although the GK rules use significantly more abscissae than the AE and Smolyak-AE rules, the resulting percentage deviation from the MAD_C is higher for $m = 5$ to $m = 9$. The use of the self-developed rules therefore leads to a faster convergence to the MAD_C what implicates that the approximations to the Gaussian integrals, which are calculated in every iteration of the filter, are of higher accuracy.

---

[1]
$$\mu_j := \mu_{j,t|t}, \ t = 1, 2, \ldots, T$$
$$y_j := y_{j,t}, \ t = 1, 2, \ldots, T$$

Table 5.1 Comparison of MADs (univariate non-stationary growth model)

	MAD$_C$	
$m = 17$ (SF=1)	6.00	
	MAD	$\Delta\%$
$m = 3$ (CKF)	7.07	17.83%
	AE/Smolyak-AE	
	MAD	$\Delta\%$
$m = 5$ (AE)	5.74	4.33%
$m = 7$ (AE)	5.85	2.50%
$m = 9$ (Smolyak-AE)	6.15	2.50%
$m = 11$ (Smolyak-AE)	6.00	0.00%
	GK	
	MAD	$\Delta\%$
$m = 5$	6.37	6.17%
$m = 7$	6.39	6.50%
$m = 9$	6.64	10.67%
$m = 11$	6.00	0.00%

Table 5.2 Comparison of computation times (univariate non-stationary growth model)

	n	∅ Time (s)	
$m = 3$ (CKF)	10	0.0283	
	AE/Smolyak-AE		
	n	∅ Time (s)	
$m = 5$ (AE)	32	0.0470	
$m = 7$ (AE)	83	0.0565	
$m = 9$ (Smolyak-AE)	361	0.0626	
$m = 11$ (Smolyak-AE)	753	0.0890	
	Smolyak		
	n	∅ Time (s)	$\Delta\%$
$m = 5$	51	0.0474	0.85%
$m = 7$	151	0.0994	75.93%
$m = 9$	401	0.1461	133.89%
$m = 11$	993	0.1122	26.07%

The average times per filter run for the tested cubature rules are shown in Table 5.2. Generally it can be stated that the filters equipped with the AE and Smolyak-AE rules work faster than those using the GK rules. The biggest difference occurs for $m = 9$, where the use of the GK rule leads to an operation time which is 133.89% of the time used by the filter based on the Smolyak-AE rule of the same exactness (column "$\Delta\%$").

At the first glance it seems self-evident that the use of less cubature abscissae must necessarily lead to a speed-up of the filter algorithm. However, as can be seen from the computation times of the filter equipped with the GK rules of exactness $m = 9$ and $m = 11$ that using the rule of exactness $m = 9$ results in a higher average computation time compared to the rule of exactness $m = 11$, although the latter rule uses more abscissae. Thus, not only the number of used abscissae is decisive for the filtering speed. With regard to the underlying state-space model, the GK rule of exactness $m = 9$ appears to generate strong instabilities which is reflected by the fact that the covariance matrix has to be corrected disproportionately often by applying Algorithm 6. To further quantify the instability, for this specific case the average number of executions of the while-slope of Algorithm 6 per filter run has been calculated. It shows that the while-slope is executed averagely 3,227 times per filter run when using the GK rule of exactness $m = 9$ and only 395 times when using the GK rule of exactness $m = 11$. In comparison, the while-slope is executed averagely zero times when using the Smolyak-AE rule of exactness $m = 9$. This indicates that reducing the influence of negative weights has a strongly positive effect on the filter stability.

5.2 The Six-Dimensional Coordinated Turn Model

The dynamics of an aircraft, executing a turn manoeuvre in the horizontal plane with constant angular speed ω, measured in rad/s, can be described by the linear system of differential equations (cf. Arasaratnam et al. 2010)

$$d\mathbf{y}_t = \mathbf{A}\mathbf{y}_t dt + \mathbf{G} d\mathbf{W}_t, \tag{5.8}$$

with

$$\mathbf{y}_t = \begin{bmatrix} \eta_t & \dot{\eta}_t & \upsilon_t & \dot{\upsilon}_t & \xi_t & \dot{\xi}_t \end{bmatrix}', \tag{5.9}$$

$$\mathbf{A} = \begin{bmatrix} 0 & 1 & 0 & 0 & 0 & 0 \\ 0 & 0 & 0 & -\omega & 0 & 0 \\ 0 & 0 & 0 & 1 & 0 & 0 \\ 0 & \omega & 0 & 0 & 0 & 0 \\ 0 & 0 & 0 & 0 & 0 & 1 \\ 0 & 0 & 0 & 0 & 0 & 0 \end{bmatrix} \tag{5.10}$$

and

$$\mathbf{G} = \mathrm{diag}\begin{bmatrix} 0 & \sigma & 0 & \sigma & 0 & \sigma \end{bmatrix}'. \tag{5.11}$$

The variables η_t, υ_t and ξ_t denote the position of the aircraft, measured in Cartesian coordinates, while $\dot{\eta}_t$, $\dot{\upsilon}_t$ and $\dot{\xi}_t$ represent the corresponding velocities. Measurements of the position in Cartesian coordinates are not available. Instead, in the present scenario a radar device is positioned at the origin which measures the range (z_{1,t_i}), the azimuth angle (z_{2,t_i}) and the aviation angle (z_{3,t_i}). Furthermore, the measurements are superimposed by the measurement noise δ_{t_i}. Thus, the vector of measurements z_{t_i} reads

$$z_{t_i} = \begin{bmatrix} \sqrt{\eta_{t_i}^2 + v_{t_i}^2 + \xi_{t_i}^2} \\ \tan^{-1}\left(\frac{v_{t_i}}{\eta_{t_i}}\right) \\ \tan^{-1}\left(\frac{\xi_{t_i}}{\sqrt{\eta_{t_i}^2 + v_{t_i}^2}}\right) \end{bmatrix} + \delta_{t_i}. \tag{5.12}$$

In order to solve the linear system (5.8), first a solution for the homogenous case $dy_t = Ay_t dt$ must be calculated:

$$\int_{t_0}^{t} \frac{1}{y_s} dy_s = A \int_{t_0}^{t} ds$$

$$\Rightarrow \log\left(\frac{y_t}{y_{t_0}}\right) = A(t - t_0) \tag{5.13}$$

$$\Rightarrow y_t = e^{A(t-t_0)} y_{t_0}.$$

Using the variation of parameters approach (cf. Zwillinger 1998, pp. 378–281),

$$y_t = e^{A(t-t_0)} c_t, \tag{5.14}$$

yields

$$\begin{aligned} dy_t &= A e^{A(t-t_0)} c_t dt + e^{A(t-t_0)} dc_t \\ &= A y_t dt + e^{A(t-t_0)} dc_t. \end{aligned} \tag{5.15}$$

Thus,

$$e^{A(t-t_0)} dc_t = G dW_t \tag{5.16}$$

$$\Rightarrow c_t = c_{t_0} + \int_{t_0}^{t} e^{-A(s-t_0)} G dW_s. \tag{5.17}$$

After inserting c_t into (5.14), the solution to (5.8) reads (cf. Arnold 1974, p. 130)

$$y_t = e^{A(t-t_0)} c_{t_0} + \int_{t_0}^{t} e^{A(t-s)} G dW_s, \quad c_{t_0} = y_{t_0}, \tag{5.18}$$

5.2 The Six-Dimensional Coordinated Turn Model

which in short notation can be expressed as

$$y_t = A^* y_{t_0} + \epsilon_t, \tag{5.19}$$

in which

$$A^* = e^{A(t-t_0)} \tag{5.20}$$

and

$$\mathbb{V}[\epsilon_t] = \int_{t_0}^{t} e^{A(t-s)} GG' e^{A'(t-s)} ds = \int_{0}^{t-t_0} e^{Au} \Omega e^{A'u} du = \Omega^*. \tag{5.21}$$

Both of the components A^* and Ω^* can be evaluated simultaneously using the method of van Loan (1978). Starting with the expression

$$B = \begin{bmatrix} -A & \Omega \\ 0 & A' \end{bmatrix}, \tag{5.22}$$

the exponential e^{Bt} yields a matrix which has the structure

$$F = \begin{bmatrix} F_{1,t} & F_{2,t} \\ 0 & F_{3,t} \end{bmatrix}. \tag{5.23}$$

From the derivative of this matrix exponential,

$$\frac{de^{Bt}}{dt} = Be^{Bt}$$
$$= \dot{F} = BF, \tag{5.24}$$

three linear differential equations with respect to the sub-matrices of F result

$$\dot{F}_{1,t} = -AF_{1,t} \tag{5.25}$$

$$\dot{F}_{2,t} = -AF_{2,t} + \Omega F_{3,t} \tag{5.26}$$

$$\dot{F}_{3,t} = A' F_{3,t}. \tag{5.27}$$

The solution to equation (5.27) reads

$$F_{3,t} = e^{A'(t-t_0)} F_{3,t_0}, \quad F_{3,t_0} = I$$
$$\Rightarrow A^* = F'_{3,t}. \tag{5.28}$$

Using this solution in (5.26) it follows that

$$\dot{F}_{2,t} = -AF_{2,t} + e^{A'(t-t_0)}\Omega. \tag{5.29}$$

The solution to (5.29) can be derived analogously to (5.13)–(5.18) and reads

$$F_{2,t} = e^{-A(t-t_0)} \int_{t_0}^{t} e^{A(s-t_0)} \Omega e^{A'(s-t_0)} ds = e^{-A(t-t_0)} \int_{0}^{t-t_0} e^{Au} \Omega e^{A'u} du \tag{5.30}$$

and thus, $\Omega^* = F'_{3,t} F_{2,t}$.

Finally, (5.18) can be written recursively in terms of discrete, not necessarily equidistant, measurement times t_i ($\Delta t = t_{i+1} - t_i$). The expression reads

$$y_{t_{i+1}} = e^{A\Delta t} y_{t_i} + \int_{0}^{\Delta t} e^{A(\Delta t - s)} G dW_{t_i + s} \tag{5.31}$$

and is called exact discrete model (EDM) (cf. Bergstrom 1966). Calculating A^* and Ω^* with the method of van Loan yields the matrices

$$A^* = \begin{bmatrix} 1 & \frac{\sin(\omega \Delta t)}{\omega} & 0 & \frac{\cos(\omega \Delta t)-1}{\omega} & 0 & 0 \\ 0 & \cos(\omega \Delta t) & 0 & -\sin(\omega \Delta t) & 0 & 0 \\ 0 & \frac{2\sin\left(\frac{\omega \Delta t}{2}\right)^2}{\omega} & 1 & \frac{\sin(\omega \Delta t)}{\omega} & 0 & 0 \\ 0 & \sin(\omega \Delta t) & 0 & \cos(\omega \Delta t) & 0 & 0 \\ 0 & 0 & 0 & 0 & 1 & \Delta t \\ 0 & 0 & 0 & 0 & 0 & 1 \end{bmatrix} \tag{5.32}$$

and

$$\Omega^* = \begin{bmatrix} -\frac{(\sin(\omega \Delta t)-\omega \Delta t)}{\omega^3/(2\sigma^2)} & \frac{\sin\left(\frac{\omega \Delta t}{2}\right)^2}{\omega^2/(2\sigma^2)} & 0 & -\frac{(\sin(\omega \Delta t)-\omega \Delta t)}{\omega^2/\sigma^2} & 0 & 0 \\ \frac{\sin\left(\frac{\omega \Delta t}{2}\right)^2}{\omega^2/(2\sigma^2)} & \sigma^2 \Delta t & \frac{(\sin(\omega \Delta t)-\omega \Delta t)}{\omega^2/\sigma^2} & 0 & 0 & 0 \\ 0 & \frac{(\sin(\omega \Delta t)-\omega \Delta t)}{\omega^2/\sigma^2} & -\frac{(\sin(\omega \Delta t)-\omega \Delta t)}{(\omega^3)/(2\sigma^2)} & \frac{\sin\left(\frac{\omega \Delta t}{2}\right)^2 \sigma^2}{(\omega^2)/2} & 0 & 0 \\ -\frac{(\sin(\omega \Delta t)-\omega \Delta t)}{\omega^2/\sigma^2} & 0 & \frac{\sin\left(\frac{\omega \Delta t}{2}\right)^2}{(\omega^2)/(2\sigma^2)} & \sigma^2 \Delta t & 0 & 0 \\ 0 & 0 & 0 & 0 & \frac{\Delta t^3 \sigma^2}{3} & \frac{\Delta t^2 \sigma^2}{2} \\ 0 & 0 & 0 & 0 & \frac{\Delta t^2 \sigma^2}{2} & \sigma^2 \Delta t \end{bmatrix}. \tag{5.33}$$

The objective of the simulation study of this section is to filter the coordinated turn model and the turn parameter ω. In order to achieve this, the parameter has to be included in the state. The dimension of resulting model therefore increases from six to seven. To avoid the time-consuming integration

5.2 The Six-Dimensional Coordinated Turn Model

over Ω^* a Taylor series expansion of Ω^* instead of the exact matrix can be used, in which ω does not appear. This approximation then will be extended appropriately in order to be used within the final seven-dimensional state-space model. The third order Taylor series expansion of Ω^* about the point $a = 0$ reads

$$T_3\left[\Omega^*\right] = \begin{bmatrix} 0 & \frac{\sigma^2 \Delta t^2}{2} & 0 & 0 & 0 & 0 \\ \frac{\sigma^2 \Delta t^2}{2} & \sigma^2 \Delta t & 0 & 0 & 0 & 0 \\ 0 & 0 & 0 & \frac{\sigma^2 \Delta t^2}{2} & 0 & 0 \\ 0 & 0 & \frac{\sigma^2 \Delta t^2}{2} & \sigma^2 \Delta t & 0 & 0 \\ 0 & 0 & 0 & 0 & 0 & \frac{\sigma^2 \Delta t^2}{2} \\ 0 & 0 & 0 & 0 & \frac{\sigma^2 \Delta t^2}{2} & \sigma^2 \Delta t \end{bmatrix}. \qquad (5.34)$$

Unfortunately, this matrix is not positive definite and therefore unusable, which leads to the fourth order Taylor series expansion of Ω^* about $a = 0$

$$T_4\left[\Omega^*\right] = \begin{bmatrix} \frac{\sigma^2 \Delta t^3}{3} & \frac{\sigma^2 \Delta t^2}{2} - \frac{\omega^2 \sigma^2 \Delta t^4}{24} & 0 & \frac{\omega \sigma^2 \Delta t^3}{6} & 0 & 0 \\ \frac{\sigma^2 \Delta t^2}{2} - \frac{\omega^2 \sigma^2 \Delta t^4}{24} & \sigma^2 \Delta t & -\frac{\omega \sigma^2 \Delta t^3}{6} & 0 & 0 & 0 \\ 0 & -\frac{\omega \sigma^2 \Delta t^3}{6} & \frac{\sigma^2 \Delta t^3}{3} & \frac{\sigma^2 \Delta t^2}{2} - \frac{\omega^2 \sigma^2 \Delta t^4}{24} & 0 & 0 \\ \frac{\omega \sigma^2 \Delta t^3}{6} & 0 & \frac{\sigma^2 \Delta t^2}{2} - \frac{\omega^2 \sigma^2 \Delta t^4}{24} & \sigma^2 \Delta t & 0 & 0 \\ 0 & 0 & 0 & 0 & \frac{\sigma^2 \Delta t^3}{3} & \frac{\sigma^2 \Delta t^2}{2} \\ 0 & 0 & 0 & 0 & \frac{\sigma^2 \Delta t^2}{2} & \sigma^2 \Delta t \end{bmatrix}. \qquad (5.35)$$

Discarding all terms containing ω for reasons of further simplification leads to a covariance matrix which has been analogously used by Arasaratnam and Haykin (2009) for the four-dimensional case and which will also be used in the upcoming simulation study:

$$\widetilde{\Omega^*} = \begin{bmatrix} \frac{\sigma^2 \Delta t^3}{3} & \frac{\sigma^2 \Delta t^2}{2} & 0 & 0 & 0 & 0 \\ \frac{\sigma^2 \Delta t^2}{2} & \sigma^2 \Delta t & 0 & 0 & 0 & 0 \\ 0 & 0 & \frac{\sigma^2 \Delta t^3}{3} & \frac{\sigma^2 \Delta t^2}{2} & 0 & 0 \\ 0 & 0 & \frac{\sigma^2 \Delta t^2}{2} & \sigma^2 \Delta t & 0 & 0 \\ 0 & 0 & 0 & 0 & \frac{\sigma^2 \Delta t^3}{3} & \frac{\sigma^2 \Delta t^2}{2} \\ 0 & 0 & 0 & 0 & \frac{\sigma^2 \Delta t^2}{2} & \sigma^2 \Delta t \end{bmatrix}. \qquad (5.36)$$

After extending the state vector and the covariance matrix by one dimension, the final state-space model used in the simulation study and its parameters read

$$\begin{bmatrix} \eta_i \\ \dot{\eta}_i \\ v_i \\ \dot{v}_i \\ \xi_i \\ \dot{\xi}_i \\ \omega_i \end{bmatrix} = \begin{bmatrix} 1 & \frac{\sin(\omega_{i-1}\Delta t)}{\omega_{i-1}} & 0 & \frac{\cos(\omega_{i-1}\Delta t)-1}{\omega_{i-1}} & 0 & 0 & 0 \\ 0 & \cos(\omega_{i-1}\Delta t) & 0 & -\sin(\omega_{i-1}\Delta t) & 0 & 0 & 0 \\ 0 & \frac{2\sin\left(\frac{\omega_{i-1}\Delta t}{2}\right)^2}{\omega_{i-1}} & 1 & \frac{\sin(\omega_{i-1}\Delta t)}{\omega_{i-1}} & 0 & 0 & 0 \\ 0 & \sin(\omega_{i-1}\Delta t) & 0 & \cos(\omega_{i-1}\Delta t) & 0 & 0 & 0 \\ 0 & 0 & 0 & 0 & 1 & \Delta t & 0 \\ 0 & 0 & 0 & 0 & 0 & 1 & 0 \\ 0 & 0 & 0 & 0 & 0 & 0 & 1 \end{bmatrix} \begin{bmatrix} \eta_{i-1} \\ \dot{\eta}_{i-1} \\ v_{i-1} \\ \dot{v}_{i-1} \\ \xi_{i-1} \\ \dot{\xi}_{i-1} \\ \omega_{i-1} \end{bmatrix} + g\epsilon_i \quad (5.37)$$

$$z_i = \begin{bmatrix} \sqrt{\eta_i^2 + v_i^2 + \xi_i^2} \\ \tan^{-1}\left(\frac{v_i}{\eta_i}\right) \\ \tan^{-1}\left(\frac{\xi_i}{\sqrt{\eta_i^2 + v_i^2}}\right) \end{bmatrix} + \delta_i,$$

with

$$\widetilde{\Omega}_7^* = gg' = \begin{bmatrix} \widetilde{\Omega}^* & 0 \\ 0 & \sigma_\omega^2 \end{bmatrix}, \quad (5.38)$$

$$R = \begin{bmatrix} \sigma_{z_1}^2 & 0 & 0 \\ 0 & \sigma_{z_2}^2 & 0 \\ 0 & 0 & \sigma_{z_3}^2 \end{bmatrix}, \quad (5.39)$$

$\sigma = \sqrt{0.2}\frac{m}{s}$, $\sigma_\omega = 0.007\frac{\text{rad}}{s}$, $\sigma_{z_1} = 50m$, $\sigma_{z_2} = 0.0017\text{rad}$ and $\sigma_{z_3} = 0.0017\text{rad}$. The time interval is set to $\Delta t = 1$.[2]

In order to avoid confusion, both system model and measurement model are indexed by i. The index does not represent the continuous but the discrete time domain and is, by an abuse of notation, to be interpreted as $y_i = y_{t_i}$. That means that the system model is treated as exact discrete model in the sense of Bergstrom (1966).[3] An analogous state-space model has already proven to work properly with regard to the four-dimensional coordinated turn model (cf. Arasaratnam and Haykin 2009; Jia et al. 2013). Therefore, the use of (5.37) represents a logical extension to the models previously used in literature.

---

[2]In the simulation of the state-space model, the variance σ_ω is set to zero because, as mentioned, the angular speed ω is supposed to be constant. Although from a theoretical point of view the variance of ω is zero, setting σ_ω to a small value during the filtering process can help to prevent numerical instabilities. Because of this reason and especially in order to be consistent with respect to the state-space models previously used in literature, σ_ω is given a moderate value (0.007) within the filter algorithm.

[3]Of course, this equalization is made for reasons of simplification and is mathematically not entirely correct because the extension of the state vector by ω_i leads to a nonlinear model.

5.2 The Six-Dimensional Coordinated Turn Model

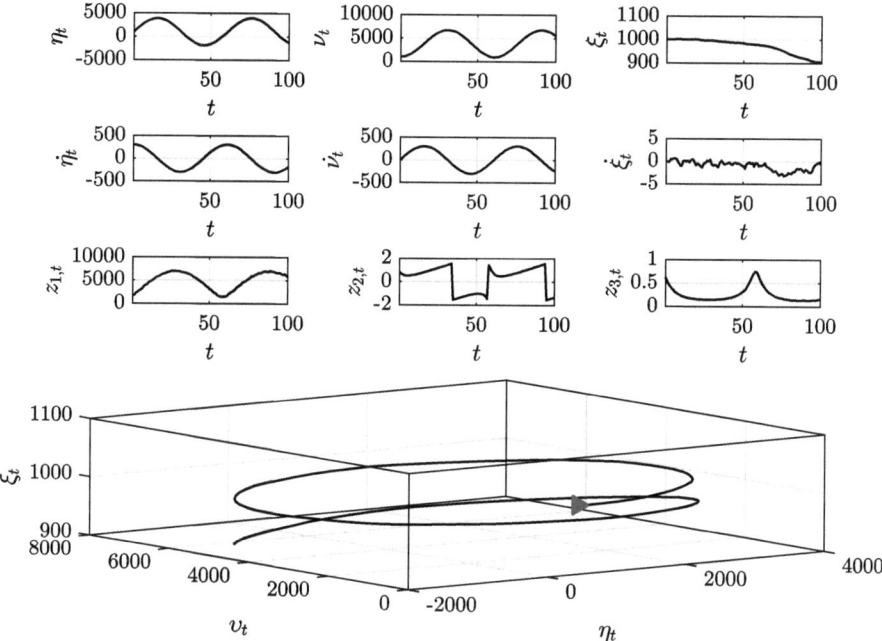

Fig. 5.4 The coordinated turn model

Figure 5.4 shows a trajectory of the coordinated turn model and the time series of position and velocity. Furthermore, also the time series of measurements are depicted. The angular speed ω in this example is $\omega = 0.1047\frac{\text{rad}}{s}$ ($\approx 6°s^{-1}$). All other parameters have the same values as in the simulation study.

The initial values used for the simulations will be chosen to be,

$$y_1 = \begin{bmatrix} 1{,}000m & 300\frac{m}{s} & 1{,}000m & 0\frac{m}{s} & 1{,}000m & 0\frac{m}{s} & \omega\frac{\text{rad}}{s} \end{bmatrix} \tag{5.40}$$

and

$$\Sigma_{yy,1|0} = \begin{bmatrix} 100m^2 & 0 & 0 & 0 & 0 & 0 & 0 \\ 0 & 10\frac{m^2}{s^2} & 0 & 0 & 0 & 0 & 0 \\ 0 & 0 & 100\,m^2 & 0 & 0 & 0 & 0 \\ 0 & 0 & 0 & 10\frac{m^2}{s^2} & 0 & 0 & 0 \\ 0 & 0 & 0 & 0 & 100m^2 & 0 & 0 \\ 0 & 0 & 0 & 0 & 0 & 10\frac{m^2}{s^2} & 0 \\ 0 & 0 & 0 & 0 & 0 & 0 & 0.25\frac{\text{rad}^2}{s^2} \end{bmatrix}. \tag{5.41}$$

$K = 500{,}000$ replications of the state-space model (5.37) with $T = 100$ will be filtered using the AE/Smolyak-AE and GK rules of exactness $m = 5$, $m = 7$,

$m = 9$ and $m = 11$ and the CKF. Based on the results, the MADs of position, velocity and angular speed ω are then calculated separately. As an example and analogous to (5.7), the formula for the MAD of the filtered position states reads

$$\text{MAD} = \frac{1}{K} \sum_{k=1}^{K} \frac{\frac{1}{T} \sum_{j=1,3,5} \left\| y_j^k - \mu_j^k \right\|_1}{3}. \tag{5.42}$$

For the approximation of the converged MADs, again a stabilized(1) rule of exactness $m = 17$ will be utilized which has a comparatively very high stability of SF = 4.56 and uses $n = 108,545$ abscissae. In each simulation run, the initial prior state estimate $\mu_{1|0}$ will be randomly drawn from $\mathcal{N}(y_1, \Sigma_{yy,1|0})$ and ω is randomly chosen from $U(-0.14, 0.14)$.

Table 5.3 shows the MADs and their absolute percentage deviation from the associated MAD$_C$ ($\Delta\%$). As to be seen in the first column, the results regarding the position are not uniform. The MADs produced by the AE/Smolyak-AE rules are closer to MAD$_C$ than those produced by the GK rules for $m = 5$ and $m = 9$ and the opposite is the case for $m = 7$ and $m = 11$. While the differences in performance between the AE/Smolyak-AE and the GK rules are moderate for $m = 7$ to $m = 11$, a notable discrepancy can be recognized with respect to rules of exactness $m = 5$. The MAD based on this Smolyak rule is extremely distant from the MAD$_C$ in comparison to the one based on the Smolyak-AE rule. Even the CKF delivers a much better performance. This indicates numerical instabilities.

Table 5.3 Comparison of MADs (coordinated turn model)

	Position		Velocity		Parameter	
	MAD$_C$		MAD$_C$		MAD$_C$	
$m = 17$ (SF=1)	10.98		4.44		0.009336	
	MAD	$\Delta\%$	MAD	$\Delta\%$	MAD	$\Delta\%$
$m = 3$ (CKF)	11.87	8.11%	9.20	107.21%	0.010666	14.24%
	AE/Smolyak-AE		AE/Smolyak-AE		AE/Smolyak-AE	
	MAD	$\Delta\%$	MAD	$\Delta\%$	MAD	$\Delta\%$
$m = 5$ (AE)	11.69	6.47%	11.45	157.88%	0.009674	3.62%
$m = 7$ (AE)	10.86	1.09%	3.53	20.50%	0.009283	0.57%
$m = 9$ (Smolyak-AE)	10.94	0.36%	4.21	5.18%	0.009310	0.28%
$m = 11$ (Smolyak-AE)	10.88	0.91%	3.71	16.44%	0.009266	0.75%
	GK		GK		GK	
	MAD	$\Delta\%$	MAD	$\Delta\%$	MAD	$\Delta\%$
$m = 5$	20.20	83.97%	90.65	1941.67%	0.013395	43.48%
$m = 7$	10.94	0.36%	4.02	9.46%	0.009239	1.04%
$m = 9$	11.11	1.18%	5.55	25.00%	0.009313	0.25%
$m = 11$	10.91	0.64%	3.89	12.39%	0.009290	0.49%

5.2 The Six-Dimensional Coordinated Turn Model

Similar results can be observed with respect to velocity. The MADs produced by the GK rules of exactness $m = 7$ and $m = 11$ are closer to the MAD_C than those produced by the AE/Smolyak-AE rules of the same exactness. For $m = 5$ and $m = 9$, the AE/Smolyak-AE rules deliver better results. The use of the GK rule of exactness $m = 5$ leads to a notably bad MAD which deviates $1{,}941.67\%$ in absolute value from the MAD_C.

Apart from the fact that the GK rule of exactness $m = 5$ again is outperformed by the CKF, the filter results regarding the angular speed ω are slightly different. The AE rules of exactness $m = 5$ and $m = 7$ outperform the corresponding GK rules, while the accuracy of the Smolyak-AE rules in comparison is lower for $m = 9$ and $m = 11$. Furthermore it can be stated that from $m = 7$ upwards all cubature rules produce MADs which are very close to the MAD_C.

The comparison of computation times is given in Table 5.4. The AE/Smolyak-AE rules use less computational resources than the GK rules. For $m = 5$ and $m = 9$ the difference is only small and the use of the GK rules leads to 2.88% and 0.12% more time on calculation compared to the corresponding AE/Smolyak-AE rules. A significant time reduction becomes apparent with regard to the rules of exactness $m = 7$ and $m = 11$. Here, the GK rules use significantly more abscissae than the AE/Smolyak-rules, which is reflected by a considerably high difference in computation time of 19.70% and 18.71%.

Summarized, the use of AE/Smolyak-AE rules of exactness $m = 5$ and $m = 7$ leads to a slight advantage with respect to computation time and, in two out of three cases, to far better filter results compared to the GK rules. The use of the AE/Smolyak-AE rules of $m = 7$ and $m = 11$ leads to a notable speed-up. But the filter results cannot keep up with the results of the GK rules in two out of three cases. It is very likely that the reason for the lack in accuracy is due to the high differences in used abscissae. As an example, the AE rule of exactness $m = 7$ uses 208 abscissae while the corresponding GK rule uses 407 abscissae, which is

Table 5.4 Comparison of computation times (coordinated turn model)

	n	⌀ Time (s)	
$m = 3$ (CKF)	14	0.04969	
AE/Smolyak-AE			
	n	⌀ Time (s)	
$m = 5$ (AE)	57	0.0598	
$m = 7$ (AE)	208	0.0710	
$m = 9$ (Smolyak-AE)	1,238	0.1414	
$m = 11$ (Smolyak-AE)	3,034	0.2470	
GK			
	n	⌀ Time (s)	$\Delta\%$
$m = 5$	99	0.0616	2.88%
$m = 7$	407	0.0850	19.70%
$m = 9$	1,317	0.1416	0.12%
$m = 11$	3,837	0.2933	18.71%

nearly twice as much. A cubature rule which possess the same degree of exactness as another, but uses significantly more abscissae, almost certainly will lead to better integration results if the function to be integrated is not close to a polynomial of appropriate degree. This is the case for the system function of the state-space model and therefore the GK rules of degree $m = 7$ and $m = 11$ have an advantage, simply because the functions to be integrated are evaluated at far more different locations. Taking into account this fact it can be concluded that the AE/Smolyak-AE rules of exactness $m = 7$ and $m = 11$ produce filter results which are comparable in accuracy to those of the GK rules.

5.3 The Lorenz Model

The Lorenz model (Lorenz 1963) is the first and probably most famous example of a system of deterministic nonlinear differential equations whose numerical solutions can exhibit chaotic behaviour. Lorenz found out that, depending on the choice of parameters, slightly different initial conditions lead to vastly different trajectories of the model. Due to its exceptional properties, the Lorenz model represents the starting point for what is today known as chaos theory.

Since the 1980s, dealing with chaos and nonlinearity in financial and macroeconomic data has become increasingly popular (cf. Faggini and Parziale 2012, p. 5). One of the hardest tasks in this field of research is to identify and prove the existence of chaotic motion within a financial or macroeconomic time series and, more specifically, to distinguish between deterministic chaos and randomness. To address this problem, various statistical test routines have been developed (cf. Matilla-García and Marín 2010). In recent decades, a multitude of different time-dependent economic phenomena has been investigated for chaotic structures and modeled by chaotic dynamical systems. These include, for example, the dynamics of exchange rates (cf. Brooks 1998; Guégan and Mercier 2006; Vlad et al. 2010), business cycles (cf. Lorenz 1987; Hallegatte et al. 2008) and stock markets (cf. Shaffer 1991).

Examinations in the filtering context of extensions to the Lorenz model by the incorporation of stochastic perturbations and a measurement equation have, inter alia, been conducted by Singer (Singer 1990, 1999, pp. 164–174) and Evensen (1997). The state-space model

$$d \begin{bmatrix} \eta_t \\ \upsilon_t \\ \xi_t \end{bmatrix} = \begin{bmatrix} -\phi_1 \eta_t + \phi_1 \upsilon_t \\ -\eta_t \xi_t + \phi_2 \eta_t - \upsilon_t \\ \eta_t \upsilon_t - \phi_3 \xi_t \end{bmatrix} dt + g d W$$

$$z_{t_i} = \begin{bmatrix} \eta_{t_i} \\ \upsilon_{t_i} \\ \xi_{t_i} \end{bmatrix} + \boldsymbol{\delta}_{t_i}$$

(5.43)

5.3 The Lorenz Model

describes the evolution of the state $y_t = \begin{bmatrix} \eta_t & \upsilon_t & \xi_t \end{bmatrix}$ in the stochastic Lorenz model. As usual, measurements are only available at discrete times $t_1, t_2, \ldots, t_N$. By the Euler method, the discretized system model reads

$$\begin{bmatrix} \eta_{t+\Delta t} \\ \upsilon_{t+\Delta t} \\ \xi_{t+\Delta t} \end{bmatrix} = \begin{bmatrix} \eta_t \\ \upsilon_t \\ \xi_t \end{bmatrix} + \begin{bmatrix} -\phi_1 \eta_t + \phi_1 \upsilon_t \\ -\eta_t \xi_t + \phi_2 \eta_t - \upsilon_t \\ \eta_t \upsilon_t - \phi_3 \xi_t \end{bmatrix} \Delta t + g\sqrt{\Delta t}\epsilon_t. \tag{5.44}$$

Figure 5.5 shows a simulation of the system model of (5.44) with the parameters $\phi_1 = 10$, $\phi_2 = 28$, $\phi_3 = \frac{8}{3}$, $\sigma_1 = 5$, $\sigma_2 = 4$, $\sigma_3 = 3$,

$$g = \begin{bmatrix} 5 & 0 & 0 \\ 0 & 4 & 0 \\ 0 & 0 & 3 \end{bmatrix}, \tag{5.45}$$

$$R = \begin{bmatrix} 0.001 & 0 & 0 \\ 0 & 0.001 & 0 \\ 0 & 0 & 0.001 \end{bmatrix} \tag{5.46}$$

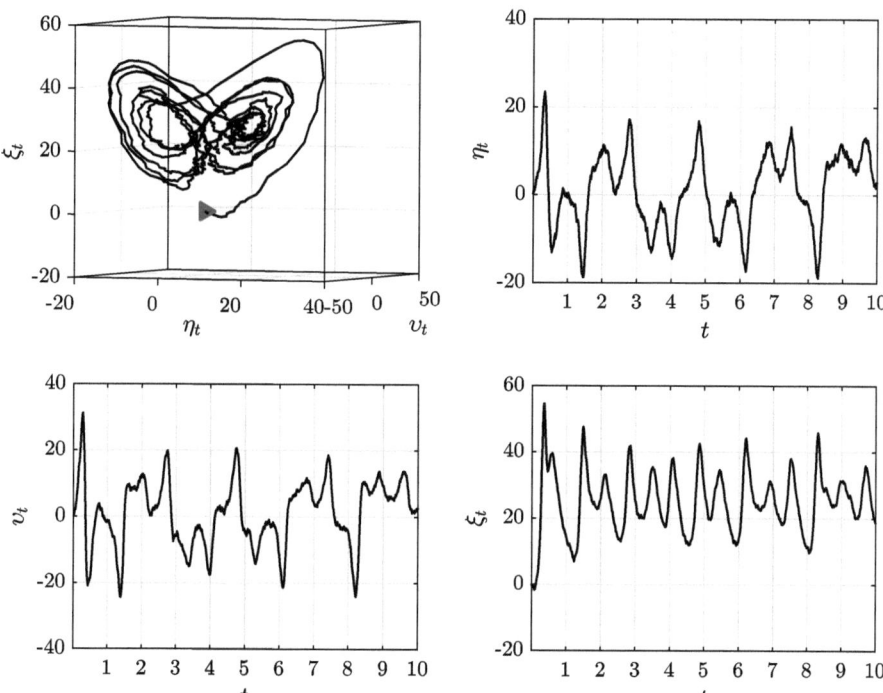

Fig. 5.5 The stochastic Lorenz model

and $\Delta t = 0.01$. This parametrization will also be used in the simulation study of this section, in which the Lorenz model and, in addition, all parameters of the system model will be filtered using the conditional Kalman filter (Algorithm 5).

The nine-dimensional state-space model reads

$$\begin{bmatrix} \eta_t \\ \upsilon_t \\ \xi_t \\ \phi_{1,t} \\ \phi_{2,t} \\ \phi_{3,t} \\ \sigma_{1,t} \\ \sigma_{2,t} \\ \sigma_{3,t} \end{bmatrix} = \begin{bmatrix} \eta_{t-1} \\ \upsilon_{t-1} \\ \xi_{t-1} \\ \phi_{1,t-1} \\ \phi_{2,t-1} \\ \phi_{3,t-1} \\ \sigma_{1,t-1} \\ \sigma_{2,t-1} \\ \sigma_{3,t-1} \end{bmatrix} + \begin{bmatrix} -\phi_{1,t-1}\eta_{t-1} + \phi_{1,t-1}\upsilon_{t-1} \\ -\eta_{t-1}\xi_{t-1} + \phi_{2,t-1}\eta_{t-1} - \upsilon_{t-1} \\ \eta_{t-1}\upsilon_{t-1} - \phi_{3,t-1}\xi_{t-1} \\ \phi_{1,t-1} \\ \phi_{2,t-1} \\ \phi_{3,t-1} \\ \sigma_{1,t-1} \\ \sigma_{2,t-1} \\ \sigma_{3,t-1} \end{bmatrix} \Delta t$$

$$+ \begin{bmatrix} \sigma_{1,t-1} & 0 & 0 & 0\,0\,0\,0\,0\,0 \\ 0 & \sigma_{2,t-1} & 0 & 0\,0\,0\,0\,0\,0 \\ 0 & 0 & \sigma_{3,t-1} & 0\,0\,0\,0\,0\,0 \\ 0 & 0 & 0 & 0\,0\,0\,0\,0\,0 \\ 0 & 0 & 0 & 0\,0\,0\,0\,0\,0 \\ 0 & 0 & 0 & 0\,0\,0\,0\,0\,0 \\ 0 & 0 & 0 & 0\,0\,0\,0\,0\,0 \\ 0 & 0 & 0 & 0\,0\,0\,0\,0\,0 \\ 0 & 0 & 0 & 0\,0\,0\,0\,0\,0 \end{bmatrix} \sqrt{\Delta t}\epsilon_t \quad (5.47)$$

$$z_{t_i} = \begin{bmatrix} \eta_{t_i} \\ \upsilon_{t_i} \\ \xi_{t_i} \end{bmatrix} + \delta_{t_i}.$$

The simulation study will be based on $K = 1,000$ replications of the state-space model, with the initial state $y_1 = \begin{bmatrix} 0\ 0\ 0\ 10\ 28\ \frac{8}{3}\ 5\ 4\ 3 \end{bmatrix}'$. The sampling interval and the time series length are set to $\Delta t = 0.01$ and $T = 10$ (1,000 gridpoints). For the purpose of filtering, 250 measurements z_{t_i} are available, which are equally spaced. Due to the complexity of the filter problem which is mainly caused by the high-dimensional state space, the filter algorithm will be only equipped with cubature rules of exactness $m = 5$ to $m = 9$.

To apply the conditional filter, the state vector y_t has to be partitioned into two sub-vectors, $y_{1,t}$ and $y_{2,t}$. The sub-vector $y_{1,t}$ includes the first six entries of y_t which are the three position states as well as the three drift parameter states. The second sub-vector, $y_{2,t}$, is three-dimensional and includes the diffusion states. The initial prior state covariance matrix for $y_{1,t}$ will be set to

5.3 The Lorenz Model

$$\Sigma_{y_{1,1}y_{1,1}|y_{2,1}=\chi_{2,m},z^0} = \begin{bmatrix} 1,000 & 0 & 0 & 0 & 0 & 0 \\ 0 & 1,000 & 0 & 0 & 0 & 0 \\ 0 & 0 & 1,000 & 0 & 0 & 0 \\ 0 & 0 & 0 & 1,000 & 0 & 0 \\ 0 & 0 & 0 & 0 & 1,000 & 0 \\ 0 & 0 & 0 & 0 & 0 & 1,000 \end{bmatrix}, \; m = 1, 2, \ldots, M. \tag{5.48}$$

The initial state estimates $\mu_{y_{1,1}|y_{2,1}=\chi_{2,m},z^0}$, $m = 1, 2, \ldots, M$, are chosen equal to each other and will be drawn randomly from $\mathcal{N}\left(\mathbf{0}, \Sigma_{y_{1,1}y_{1,1}|y_{2,1}=\chi_{2,m},z^0}\right)$. For the sub-vector $y_{2,t}$, the initial prior state covariance matrix will be chosen to be

$$\Sigma_{y_2 y_2, 1|0} = \begin{bmatrix} 100 & 0 & 0 \\ 0 & 100 & 0 \\ 0 & 0 & 100 \end{bmatrix} \tag{5.49}$$

and the initial prior state estimate $\mu_{y_2, 1|0}$ will be drawn from $\mathcal{N}\left(\mathbf{0}, \Sigma_{y_2 y_2, 1|0}\right)$.

Table 5.5 shows the resulting combined MADs with respect to position, the parameters of the drift function and the diffusion parameters which have been calculated analogous to formula (5.42). With regard to the MAD of the diffusion parameters, $\text{MAD}_{\text{vol.-prm.}}$, the calculation has been slightly changed to

$$\text{MAD} = \frac{1}{K} \sum_{k=1}^{K} \frac{\frac{1}{T} \sum_{j=7,8,9} \left\| y_j^k - \left| \mu_j^k \right| \right\|_1}{3}. \tag{5.50}$$

Table 5.5 Comparison of MADs (Lorenz model)

	$\text{MAD}_{\text{pos.}}$	$\text{MAD}_{\text{drift-prm.}}$	$\text{MAD}_{\text{vol.-prm.}}$
$m = 3$ (CKF)	0.33	0.59	1.04
	AE/Smolyak-AE		
	$\text{MAD}_{\text{pos.}}$	$\text{MAD}_{\text{drift-prm.}}$	$\text{MAD}_{\text{vol.-prm.}}$
$m = 5$ (AE)	0.33	0.68	0.59
$m = 7$ (AE)	0.33	0.67	0.76
$m = 9$ (Smolyak-AE)	0.33	0.64	2.24
	GK		
	$\text{MAD}_{\text{pos.}}$	$\text{MAD}_{\text{drift-prm.}}$	$\text{MAD}_{\text{vol.-prm.}}$
$m = 5$	0.33	0.63	1.73
$m = 7$	0.33	0.78	0.66
$m = 9$	0.33	0.74	2.58

Table 5.6 Comparison of computation times (Lorenz model)

	n	∅ Time (s)	
$m = 3$ (CKF)	72 (12 × 6)	2.14	
	AE/Smolyak-AE		
	n	∅ Time (s)	
$m = 5$ (AE)	572 (44 × 13)	4.22	
$m = 7$ (AE)	3,483 (129 × 27)	10.43	
$m = 9$ (Smolyak-AE)	42,560 (532 × 80)	61.14	
	GK		
	n	∅ Time (s)	Δ% Time
$m = 5$	1,387 (73 × 19)	6.30	49.29%
$m = 7$	10,023 (257 × 39)	19.10	83.13%
$m = 9$	69,657 (749 × 93)	87.06	42.39%

The absolute values of μ_j^k are subtracted to account for the fact that the filter solutions of the diffusion states may also converge to a correct, but negative value ($-\sigma^2 = \sigma^2$).

With respect to position, all cubature rules generate the same MAD which is caused by the fact that the first three entries of y_t are measurable in short intervals with very small measurement errors. Comparing the filter results connected to the drift parameters and the diffusion parameters, the AE/Smolyak-AE rules generate lower MADs in comparison to the GK rules in four out of six cases. As an interesting result, with respect to the estimation of the drift parameters the CKF performs better than all other methods and thus it again becomes apparent that the use of high degree cubature rules does not guarantee an increase of accuracy, if the real filter densities are not Gaussian or at least similar to Gaussian densities. In summary, it can be stated that, with respect to the chosen parametrization of the state-space model, the AE/Smolyak-AE rules lead to better filter results than the GK rules.

Table 5.6 shows the average computation time per filter run. The conditional filter utilizes two sets of cubature rules, because the state vector y_t has to be partitioned into two sub-vectors. The amounts of abscissae used by the individual rules, n_1 and n_2, are given in brackets and the total amount of abscissae used by the filter, $n_1 \cdot n_2$, is given before the brackets. The use of the GK rules leads, at minimum, to a 42% higher computation time compared to the Smolyak-AE rules of the same exactness. Due to the fact that $n_1 \cdot n_2$ abscissae are needed to run the filter, the computational efficiency is drastically increased by the use of the AE/Smolyak-AE rules.

5.4 The Ginzburg–Landau Model

The model of Ginzburg and Landau (1950) is a deterministic differential equation which has originally been designed for the purpose of modeling phase transitions between solid and liquid media. Elements of the associated Ginzburg–Landau

5.4 The Ginzburg–Landau Model

theory have also expanded into economic research questions. One example is the elaboration of a theoretical framework for the description of fluctuations and transitions in financial markets (cf. Savona et al. 2015).

The objective of this last simulation study is to filter a three-dimensional extension of the stochastic Ginzburg–Landau model with measurements equation,

$$dy_t = -\left(\alpha y_t + \beta y_t^3\right) dt + g dW_t$$
$$z_{t_i} = y_{t_i} + \delta_{t_i}.$$
(5.51)

using the discrete Bayes filter (Algorithm 1). Inspecting the potential of the Ginzburg–Landau model, which is the antiderivative of the drift function, it becomes clear that dynamics of the state equation are decisively determined by the values of the parameters α and β:

$$\Phi(y_t) = \alpha \frac{y_t^2}{2} + \beta \frac{y_t^4}{4}.$$
(5.52)

For negative values of α and $\beta > 0$, the potential becomes bimodal. Figure 5.6 shows the Ginzburg–Landau model with the parameters $\alpha = -1$ and $\beta = 1$. Due to the bimodality of the potential, the process jumps back and forth erratically between two regimes. Another consequence of the chosen parameters is the fact that the prior density becomes strongly non-Gaussian between measurements, if the measurement interval is large (cf. Singer 2006b, p. 11). Figure 5.7 shows how the prior density in the stochastic Ginzburg–Landau model with $\alpha = -1$, $\beta = 1$ changes between measurements. Starting with a Gaussian density, the posterior evolves into a bimodal density.

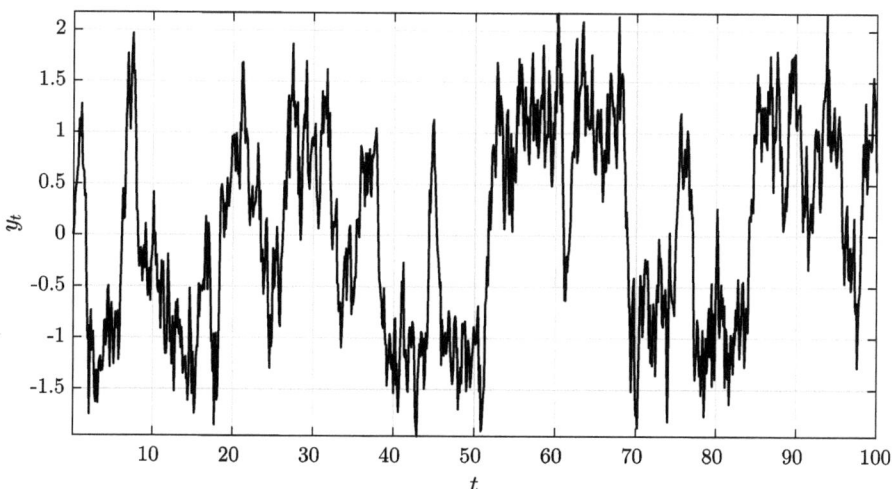

Fig. 5.6 The stochastic Ginzburg–Landau model ($\alpha = -1$, $\beta = 1$)

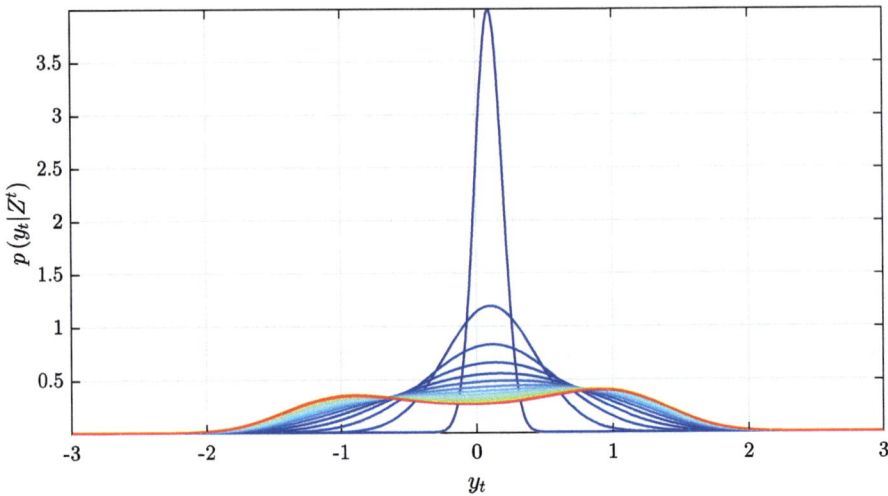

Fig. 5.7 Time evolution of the prior density between measurements ($\alpha = -1$, $\beta = 1$)

For the construction of the state-space model, the Euler discretization

$$y_{t+\Delta t} = y_t - \left(\alpha y_t + \beta y_t^3\right) \Delta t + g\sqrt{\Delta t}\epsilon_t \tag{5.53}$$

of the differential equation is replicated in three dimensions. Analogously to the first simulation study (Sect. 5.1) the three models will be coupled by correlated errors. The used state-space model reads

$$\begin{bmatrix} y_{1,t} \\ y_{2,t} \\ y_{3,t} \end{bmatrix} = \begin{bmatrix} y_{1,t-1} \\ y_{2,t-1} \\ y_{3,t-1} \end{bmatrix} - \begin{bmatrix} \alpha y_{1,t-1} + \beta y_{1,t-1}^3 \\ \alpha y_{2,t-1} + \beta y_{2,t-1}^3 \\ \alpha y_{3,t-1} + \beta y_{3,t-1}^3 \end{bmatrix} \Delta t + g\sqrt{\Delta t}\epsilon_t \tag{5.54}$$

$$z_{t_i} = y_{t_i} + \delta_{t_i}.$$

with the associated parameters

$$\Omega = gg' = \begin{bmatrix} 1 & 0.25 & -0.5 \\ 0.25 & 1 & -0.35 \\ -0.5 & -0.35 & 1 \end{bmatrix} \tag{5.55}$$

and

$$R = \begin{bmatrix} 0.5 & 0 & 0 \\ 0 & 0.5 & 0 \\ 0 & 0 & 0.5 \end{bmatrix}. \tag{5.56}$$

5.4 The Ginzburg–Landau Model

The simulated processes will be filtered using cubature rules of exactness $m = 5$ and $m = 7$. Since the discrete Bayes filter operates on the basis of the true filter densities, cubature rules for the integration of unweighted integrals must be applied. The performance of the self-created rules (Table 4.1) will compared to the performance of the DGKP rules. In order to make these rules usable within the filtering context, the rules will be extended into compound rules (Sect. 3.2.7). Additionally, also Riemann integration will be included in the simulation study.

For the simulation, $K = 15{,}000$ replications of the state-space model will be generated with the initial state $\boldsymbol{y}_1 = \begin{bmatrix} 0 & 0 & 0 \end{bmatrix}'$. The sampling interval and the time series length are set to $\Delta t = 0.1$ and $T = 100$ (1,000 gridpoints). For the filter procedure, 250 equally spaced measurements are available. The parameters α and β will be drawn randomly from $U(-5, 0)$ and $U(0, 5)$. As initial prior, a Gaussian density with $\boldsymbol{\mu} = \begin{bmatrix} 0 & 0 & 0 \end{bmatrix}'$ and

$$\boldsymbol{\Sigma} = \begin{bmatrix} 1{,}000 & 0 & 0 \\ 0 & 1{,}000 & 0 \\ 0 & 0 & 1{,}000 \end{bmatrix}. \tag{5.57}$$

will be chosen. The integrations will be carried out over the cube $[-3.5, 3.5]^d$, $d = 3$.

The results of the simulation study are given in Table 5.7. As an example to clarify the notation, the AE rule of exactness $m = 5$ uses $n = 13$ abscissae and is extended to a compound rule of level $l = 8$. Thus, the domain of integration is decomposed into $2^8 = 256$ subregions and the resulting rule uses $256 \cdot 13 = 3328$ abscissae. In the table, this combination is represented by $3{,}328/13/\text{Level } 8$.

In the case of the exact discrete filter, the real filter densities are not replaced by Gaussians. Therefore, the MADs generated by the different cubature rules are to be interpreted as direct indicators for the quality of the applied cubature rules. It turns out that the AE rules consistently generate lower MADs than the DGKP rules, although the amount of abscissae used is much lower. One reason for this

Table 5.7 Comparisons of MADs and computation times (Ginzburg–Landau model)

	Riemann			
	MAD	n	Ø Time (s)	
	0.7583	3,375/15	5.34	
	AE/Smolyak-AE			
	MAD	n	Ø Time (s)	
$m = 5$ (AE)	0.7555	3,328/13/Level 8	5.30	
$m = 7$ (AE)	0.7788	3,328/26/Level 7	5.30	
	DGKP			
	MAD	n	Ø Time (s)	$\Delta\%$ Time
$m = 5$	0.7705	4,864/19/Level 8	9.17	73.02%
$m = 7$	7.23	4,992/39/Level 7	9.40	77.36%

may be the negative weights which are used by the DGKP rules. By the extension of the cubature rules into compound rules, the negative weights are applied in every subregion of the domain of integration. Consequently, this may lead to an increased accumulation of rounding errors. The weights of the AE rules instead are strictly positive. Furthermore, lowering the level of decomposition and in exchange increasing the exactness of the basic cubature rule has negative effects. The MADs generated by the rules of exactness $m = 7$ are worse than the MADs generated by the rules of exactness $m = 5$, which indicates that the density of the cubature grid is the most important requirement in order to obtain accurate results with the exact discrete filter. This deterioration of filtering accuracy is most conspicuous with respect to the DGKP rules. An additional problem is the fact that the DGKP rule of exactness $m = 7$ is even more unstable than the DGKP rule of exactness $m = 5$. Another notable result is that Riemann integration, even though it is not an interpolatory cubature rule, performs almost as good as the AE rules. When inspecting the average time per filter run, the use of the AE rules leads to a clear superiority in terms of computational load. The filter equipped with DGKP rules in both cases is over 70% slower than the filter equipped with the AE rules.

Chapter 6
Results

The main purpose of this work was the development of new stable and efficient methods of multidimensional deterministic numerical integration (cubature rules) for the use within Bayesian filter algorithms. The following remarks are intended to give an overview of the results achieved.

In Sect. 4.1 the so-called AE rules (approximate exactness), suitable for unweighted and Gaussian Integrals, are derived. The used algorithm is based on the multidimensional moment equations and represents a least squares approach to the construction of cubature rules. The term "least squares" always implicates approximation errors and therefore the resulting rules have an *approximate* degree of polynomial exactness. However, the errors are so small that they can be described as insignificant. Due to the computational intensity of the method, AE rules are only constructed for $m = 5$, dimensions $d = 2$ to $d = 12$ and $m = 7$, dimensions $d = 2$ to $d = 8$. These rules use strictly positive weights and therefore possess optimal stability. Furthermore, for the most of the considered cases, the number of used abscissae is smaller than the number used by the most efficient rules known in literature. In all other cases, the AE rules use as many or less abscissae than the most efficient alternative rules of optimal stability.

An approach to the construction of rules of higher exactness for higher dimensions is presented in Sect. 4.2. Here, the technique used for the derivation of new cubature rules is based on the Smolyak algorithm. The first kind of new Smolyak cubature rules are the stabilized(1) rules. They are presented for the cases of unweighted as well as the Gaussian integrals and exemplary for the exactness $m = 13$ and the dimensions $d = 2$ to $d = 12$. The focus in the construction of these rules lies mainly on stability and as a result, the amount of abscissae used by the stabilized(1) rules is approximately 30% higher than the amount used by the comparison methods of Heiss and Winschel. As far as the stability is concerned, the computed rules show a significant superiority over the DGKP rules and the GK rules, proposed by Heiss and Winschel. The use of the stabilized(1) rules makes sense in situations where the speed of computation is subordinated and

© Springer International Publishing AG 2017
D. Ballreich, *Stable and Efficient Cubature-based Filtering in Dynamical Systems*, DOI 10.1007/978-3-319-62130-2_6

only the accuracy of the computed integral is crucial. Beyond the given examples, solutions can also easily be obtained with regard to higher degrees of exactness and dimensions by the use of the derived algorithm.

The second kind of self-constructed Smolyak cubature rules are the stabilized(2) rules which are constructed with special focus on the degrees of exactness $m = 9$ and $m = 11$. The rules are given exemplary for the dimensions $d = 2$ to $d = 10$. The used approach shows to work best in the case of Gaussian integrals and thus the unweighted case is not considered. For $m = 9, d > 4$ and $m = 11, d > 5$, no other rules known from literature are more efficient than the GK rules. The stabilized(2) rules represent a clear improvement for the following reasons. Firstly, the stability is much higher for both $m = 9$ and $m = 11$. Although the improvement in stability is lower than the improvement reached by the stabilized(1) rules, the difference compared to the GK rules is still very clear. Secondly, for $m = 9$ the amount of abscissae used is equal to the amount used by the GK rules. For $m = 11$, the number of abscissae used by the stabilized(2) is slightly higher in comparison to the GK rules. However, the percentage difference is almost negligible and, moreover, tends to zero with rising d. Using the proposed algorithm, also stable rules for higher dimensions can be obtained easily.

The third kind of self-constructed cubature rules are the Smolyak-AE rules which are given for the case of Gaussian integrals, degrees of exactness $m = 9$ and $m = 11$ and dimensions $d = 2$ to $d = 10$. The Smolyak-AE rules are designed with the goal to outperform the GK rules in terms of stability *and* efficiency. The applied approach is based on the idea of shrinking a percentage of the cubature weights towards zero so that the affiliated abscissae may be omitted. Even if the weights concerned are vanishingly small, omitting the corresponding abscissae entails small inaccuracies of the cubature rules. Therefore, as in the case of the already mentioned AE rules, the degrees of polynomial exactness of the Smolyak-AE rules are approximate. In comparison to the stabilized(1) and the stabilized(2) rules, the Smolyak-AE rules have a lower level of stability as a result of the attempt to increase the efficiency. Nevertheless, the Smolyak-AE rules are clearly more stable than the GK rules. As already mentioned before, for the cases $m = 9, d > 4$ and $m = 11, d > 5$ no other rules known from literature are more efficient than the GK rules. The Smolyak-AE rules on the average use significantly less abscissae than the GK rules and moreover, significantly less abscissae than the most efficient rules known for $m = 9, d \leq 4$ and $m = 11, d \leq 5$ which have been found by Stenger (cf. Stenger 1971, pp. 7–9). Ignoring the small numerical inaccuracies, the Smolyak-AE rules represent the most efficient rules for $m = 9$ and $m = 11$ for dimensions $d = 2$ to $d = 10$.

The main differences between the new kinds of Smolyak cubature rules can be summarized as follows:

- Stability ranking: Stabilized(1)>Stabilized(2)>Smolyak-AE.
- All three kinds of rules are of significantly higher stability than the methods of comparison.
- Efficiency ranking: Smolyak-AE>Stabilized(2)>Stabilized(1).

6 Results

- The stabilized(1) rules and the stabilized(2) rules are exact with respect to the integrations of polynomials while the Smolyak AE-rules have an *approximate* degree of polynomial exactness.
- The stabilized(2) rules use as many abscissae as the methods of comparison for $m = 9$ and only slightly more for $m = 11$.
- The Smolyak-AE rules are, apart from small inaccuracies with respect to the degree of polynomial exactness, more efficient than all other rules known from literature for $m = 9$ and $m = 11$ and dimensions $d = 2$ to $d = 10$.

In the fifth chapter, the performance of a selection of self-constructed cubature rules is compared to the performance of the GK and DGKP rules within five extensive simulation studies. In order to put the emphasis on efficiency *and* stability, the AE and Smolyak-AE rules are chosen for the comparison. The test investigated models are the non-stationary growth model (Kitagawa 1987), the six-dimensional coordinated turn model (cf. Arasaratnam et al. 2010), the Lorenz model (Lorenz 1963) and the Ginzburg–Landau model (Ginzburg and Landau 1950). As the results show, the self-created cubature rules can be used for the purpose of filtering without any difficulties. Although the AE and Smolyak-AE rules have only an *approximate* degree of polynomial exactness, there are no numerical problems such as divergences of certain filter algorithms to be observed. Quite the contrary is the case. In the simulation study using the non-stationary growth model it turns out that the GK rule of exactness $m = 9$ produces strong instabilities which is not the case for the Smolayk-AE rule of the same exactness. As far as the convergence properties are concerned, in three of the four conducted simulation studies the self-constructed rules show better behaviour than the GK/DGKP rules. Those are the studies using the non-stationary growth model the Lorenz model and the Ginzburg–Landau model. With regard to the study which uses the coordinated turn model, the performance of the self-created rules on the average is equal to the performance of the comparison rules.

In terms of efficiency, the AE and Smolyak-AE rules show to be superior across all simulation studies. As an example the studies connected to the Lorenz model and the Ginzburg–Landau model are to be mentioned. In the case of the Lorenz model, the use of the GK rules leads to an average time consumption per filter run which is at minimum 49.29% higher compared to the self-constructed rules. With regard to the Ginzburg–Landau model, the comparison methods at minimum generate an average time per filter run which is even 73.02% higher.

Another purpose of the conducted simulation studies was to provide an understanding of the fact that exchanging a cubature rule of low exactness used within a Gaussian filter algorithm by a rule of high exactness does not necessarily lead to an improvement of the filter results. The Gaussian assumption $p\left(y_t | Z^{t-1}\right) \approx \mathcal{N}\left(\mu_{y,t|t-1}, \Sigma_{yy,t|t-1}\right)$ and $p\left(y_t | Z^t\right) \approx \mathcal{N}\left(\mu_{y,t|t}, \Sigma_{yy,t|t}\right)$ inevitably leads to a distortion of the filter solutions and the magnitude of this distortion depends on how similar the true densities are to Gaussian densities. If the state-space model is very nonlinear, the similarities will most likely be very small. Increasing the exactness of the cubature rules will therefore cause a more exact approximation of the "wrong" integrals. Increasing the exactness even more, the filter result will converge to some solution which may or may not be close to the latent state. Inspecting, for instance, the filter performance of the cubature Kalman filter ($m = 3$) in the Lorenz model with respect to the estimation of the drift parameters, it turns out that the resulting mean absolute deviation is significantly smaller than those generated by high degree cubature rules.

Appendix A
The Conditional Mean

The conditional mean

$$\mathbb{E}\left[y_t|Z^t\right] = \int_{-\infty}^{\infty} y_t p\left(y_t|Z^t\right) dy_t \qquad (A.1)$$

is the minimum variance unbiased estimator for the latent state y_t given the measurements Z^t which can be shown as follows. Given an arbitrary unbiased estimator $g\left(Z^t\right)$ for y_t so that $\mathbb{E}\left[g\left(Z^t\right)\right] = \mathbb{E}\left[y_t\right]$ and proposing $a'\mathbb{V}\left[\mathbb{E}\left[y_t|Z^t\right]\right]a \leq a'\mathbb{V}\left[g\left(Z^t\right)\right]a$ for every non-zero column vector a, what means that the difference $\mathbb{V}\left[g\left(Z^t\right)\right] - \mathbb{V}\left[\mathbb{E}\left[y_t|Z^t\right]\right]$ is positive semidefinite, leads to

$$a'\mathbb{V}\left[\mathbb{E}\left[y_t|Z^t\right]\right]a \leq a'\mathbb{V}\left[g\left(Z^t\right)\right]a$$

$$\Leftrightarrow$$

$$a'\mathbb{E}\left[\left(\mathbb{E}\left[y_t|Z^t\right] - \mathbb{E}\left[y_t\right]\right)\left(\mathbb{E}\left[y_t|Z^t\right] - \mathbb{E}\left[y_t\right]\right)'\right]a \leq \qquad (A.2)$$
$$a'\mathbb{E}\left[\left(g\left(Z^t\right) - \mathbb{E}\left[y_t\right]\right)\left(g\left(Z^t\right) - \mathbb{E}\left[y_t\right]\right)'\right]a$$

Subtracting and adding $\mathbb{E}\left[y_t|Z^t\right]$ on the right-hand side yields

$$a'\mathbb{E}\left[\left(\mathbb{E}\left[y_t|Z^t\right] - \mathbb{E}\left[y_t\right]\right)\left(\mathbb{E}\left[y_t|Z^t\right] - \mathbb{E}\left[y_t\right]\right)'\right]a \leq$$
$$a'\mathbb{E}\left[\left(g\left(Z^t\right) - \mathbb{E}\left[y_t|Z^t\right] + \mathbb{E}\left[y_t|Z^t\right] - \mathbb{E}\left[y_t\right]\right)\right.$$
$$\left.\cdot\left(g\left(Z^t\right) - \mathbb{E}\left[y_t|Z^t\right] + \mathbb{E}\left[y_t|Z^t\right] - \mathbb{E}\left[y_t\right]\right)'\right]a$$

$$\Leftrightarrow$$

$$a'\mathbb{E}\left[\left(\mathbb{E}\left[y_t|Z^t\right] - \mathbb{E}\left[y_t\right]\right)\left(\mathbb{E}\left[y_t|Z^t\right] - \mathbb{E}\left[y_t\right]\right)'\right]a \leq \quad \text{(A.3)}$$
$$a'\left[\mathbb{E}\left[\left(g\left(Z^t\right) - \mathbb{E}\left[y_t|Z^t\right]\right)\left(g\left(Z^t\right) - \mathbb{E}\left[y_t|Z^t\right]\right)'\right]\right.$$
$$+2\mathbb{E}\left[\left(g\left(Z^t\right) - \mathbb{E}\left[y_t|Z^t\right]\right)\left(\mathbb{E}\left[y_t|Z^t\right] - \mathbb{E}\left[y_t\right]\right)'\right]$$
$$\left.+\mathbb{E}\left[\left(\mathbb{E}\left[y_t|Z^t\right] - \mathbb{E}\left[y_t\right]\right)\left(\mathbb{E}\left[y_t|Z^t\right] - \mathbb{E}\left[y_t\right]\right)'\right]\right]a'.$$

Using the law of total expectation, one can rewrite the middle expression of the right-hand side into

$$2\mathbb{E}\left[\mathbb{E}\left[\left(g\left(Z^t\right) - \mathbb{E}\left[y_t|Z^t\right]\right)\left(\mathbb{E}\left[y_t|Z^t\right] - \mathbb{E}\left[y_t\right]\right)'|Z^t\right]\right]$$

$$\Leftrightarrow$$

$$2\mathbb{E}\left[\left(\mathbb{E}\left[y_t|Z^t\right] - \mathbb{E}\left[y_t\right]\right)\mathbb{E}\left[\left(g\left(Z^t\right) - \mathbb{E}\left[y_t|Z^t\right]\right)'|Z^t\right]\right] \quad \text{(A.4)}$$

$$\Rightarrow$$

$$2\mathbb{E}\left[\left(\mathbb{E}\left[y_t|Z^t\right] - \mathbb{E}\left[y_t\right]\right)\left(\mathbb{E}\left[y_t\right] - \mathbb{E}\left[y_t\right]\right)'\right] = \mathbf{0}.$$

Thus, the inequality holds and the result is

$$a'\mathbb{E}\left[\left(\mathbb{E}\left[y_t|Z^t\right] - \mathbb{E}\left[y_t\right]\right)\left(\mathbb{E}\left[y_t|Z^t\right] - \mathbb{E}\left[y_t\right]\right)'\right]a \leq$$
$$a'\left[\mathbb{E}\left[\left(g\left(Z^t\right) - \mathbb{E}\left[y_t|Z^t\right]\right)\left(g\left(Z^t\right) - \mathbb{E}\left[y_t|Z^t\right]\right)'\right]\right. \quad \text{(A.5)}$$
$$\left.+\mathbb{E}\left[\left(\mathbb{E}\left[y_t|Z^t\right] - \mathbb{E}\left[y_t\right]\right)\left(\mathbb{E}\left[y_t|Z^t\right] - \mathbb{E}\left[y_t\right]\right)'\right]\right]a.$$

The last expression only holds with equality, if $g\left(Z^t\right) = \mathbb{E}\left[y_t|Z^t\right]$.

Appendix B
The Moments of the Conditional Normal Distribution

Let y and z be bivariate normal distributed random variables. Then the density $p(y|z)$ is the conditional normal distribution

$$p(y|z) \sim \mathcal{N}\left(\mu_y + \frac{\sigma_{yz}}{\sigma_z^2}(z-\mu_z), \sigma_y^2 - \frac{\sigma_{yz}^2}{\sigma_z^2}\right). \tag{B.1}$$

This can be derived as follows:

$$p(y|z) = \frac{\frac{1}{2\pi\sigma_y\sigma_z\sqrt{1-\rho^2}}\exp\left[-\frac{1}{2(1-\rho^2)}\left(\frac{(y-\mu_y)^2}{\sigma_y^2} - 2\rho\frac{y-\mu_y}{\sigma_y}\cdot\frac{z-\mu_z}{\sigma_z} + \frac{(z-\mu_z)^2}{\sigma_z^2}\right)\right]}{\frac{1}{\sigma_z\sqrt{2\pi}}\exp\left(-\frac{1}{2}\frac{(z-\mu_z)^2}{\sigma_z^2}\right)}$$

$$= \frac{\exp\left[-\frac{1}{2}\left(\frac{(y-\mu_y)^2}{\sigma_y^2(1-\rho^2)} - 2\rho\frac{y-\mu_y}{\sigma_y(1-\rho^2)}\cdot\frac{z-\mu_z}{\sigma_z} + \rho^2\frac{(z-\mu_z)^2}{\sigma_z^2(1-\rho^2)}\right)\right]}{\sqrt{2\pi}\sqrt{\sigma_y^2 - \frac{\sigma_{yz}^2}{\sigma_z^2}}}$$

$$= \frac{\exp\left[-\frac{1}{2}\left(\frac{(y-\mu_y)}{\sigma_y\sqrt{1-\rho^2}} - \rho\frac{(z-\mu_z)}{\sigma_z\sqrt{1-\rho^2}}\right)^2\right]}{\sqrt{2\pi}\sqrt{\sigma_y^2 - \frac{\sigma_{yz}^2}{\sigma_z^2}}}$$

$$= \frac{\exp\left[-\frac{1}{2}\left(\frac{(y-\mu_y)}{\sqrt{\sigma_y^2 - \frac{\sigma_{yz}^2}{\sigma_z^2}}} - \frac{\sigma_{yz}}{\sigma_z^2\sigma_y\sqrt{1-\rho^2}}(z-\mu_z)\right)^2\right]}{\sqrt{2\pi}\sqrt{\sigma_y^2 - \frac{\sigma_{yz}^2}{\sigma_z^2}}}$$

$$= \frac{\exp\left[-\frac{1}{2}\left(\frac{(y-\mu_y)}{\sqrt{\sigma_y^2-\frac{\sigma_{yz}^2}{\sigma_z^2}}} - \frac{\frac{\sigma_{yz}}{\sigma_z^2}(z-\mu_z)}{\sqrt{\sigma_y^2-\frac{\sigma_{yz}^2}{\sigma_z^2}}}\right)^2\right]}{\sqrt{2\pi}\sqrt{\sigma_y^2-\frac{\sigma_{yz}^2}{\sigma_z^2}}}$$

(B.2)

$$= \frac{1}{\sqrt{2\pi}\sqrt{\sigma_y^2-\frac{\sigma_{yz}^2}{\sigma_z^2}}}\exp\left[-\frac{1}{2}\left(\frac{y-\left(\mu_y+\frac{\sigma_{yz}}{\sigma_z^2}(z-\mu_z)\right)}{\sqrt{\sigma_y^2-\frac{\sigma_{yz}^2}{\sigma_z^2}}}\right)^2\right].$$

In the multivariate case, where y and z are normal distributed random vectors which are correlated with one another, the moments are written as (cf. Liptser and Shiryaev 2001, p. 61):

$$\begin{aligned}\boldsymbol{\mu}_{y|z} &= \boldsymbol{\mu}_y + \boldsymbol{\Sigma}_{yz}\boldsymbol{\Sigma}_{zz}^{-1}(z-\boldsymbol{\mu}_z) \\ \boldsymbol{\Sigma}_{y|z} &= \boldsymbol{\Sigma}_{yy} - \boldsymbol{\Sigma}_{yz}\boldsymbol{\Sigma}_{zz}^{-1}\boldsymbol{\Sigma}_{yz}'\end{aligned}$$

(B.3)

Appendix C
The Golub–Welsch Algorithm

Any set of monic orthogonal polynomials $p_l(x)$ satisfies the following recurrence relation (cf. Wilf 1962, p. 53):

$$p_{l+1}(x) = (x - a_l) p_l(x) - b_l p_{l-1}(x). \tag{C.1}$$

This can be derived as follows (cf. Gil et al. 2007, p. 138). Because of the above-mentioned monicity, the difference polynomial $p_{l+1}(x) - x p_l(x)$ has degree at most l. Therefore, it can be constructed as a weighted sum of the orthogonal polynomials of degree up to l:

$$p_{l+1}(x) - x p_l(x) = \sum_{j=0}^{l} \xi_i p_i(x). \tag{C.2}$$

Taking the scalar product with $p_j(x)$, $j \leq l$, equation (C.2) reduces to $-\langle x p_l(x), p_j(x) \rangle = \xi_j \|p_j(x)\|_2^2$. This due to two facts, which hold by construction:

$$\langle p_{l+1}(x), p_j(x) \rangle = 0, \text{ because } j \leq l \tag{C.3}$$

and

$$\langle \xi_i p_i(x), p_j(x) \rangle = 0, \text{ for } i \neq j \tag{C.4}$$

Furthermore, for $j \leq l - 2$ also $-\langle x p_l(x), p_j(x) \rangle = -\langle p_l(x), x p_j(x) \rangle = 0$, because in this case the degree of $x p_j(x)$ is always smaller than l. In conclusion, $\xi_i = 0$ for

$i \leq l-2$. For the remaining coefficients ξ_{l-1} and ξ_l, concrete formulas can be stated:

$$-\langle xp_l(x), p_{l-1}(x)\rangle = \xi_{l-1} \|p_{l-1}(x)\|_2^2 \Leftrightarrow \xi_{l-1} = -\frac{\langle p_l(x), xp_{l-1}(x)\rangle}{\|p_{l-1}(x)\|_2^2} \tag{C.5}$$

According to (C.2), $\langle p_l(x), xp_{l-1}(x)\rangle$ can be expressed as

$$\left\langle p_l(x), p_l(x) - \sum_{j=0}^{l-1} \xi_i p_i(x) \right\rangle = \|p_l(x)\|_2^2, \tag{C.6}$$

and from this it can be concluded that

$$-\xi_{l-1} = \frac{\|p_l(x)\|_2^2}{\|p_{l-1}(x)\|_2^2} = b_l. \tag{C.7}$$

For the parameter a_l one finds

$$-\langle xp_l(x), p_l(x)\rangle = \xi_l \|p_l(x)\|_2^2 \Rightarrow -\xi_l = \frac{\langle xp_l(x), p_l(x)\rangle}{\|p_l(x)\|_2^2} = a_l. \tag{C.8}$$

For the description of the Golub–Welsch algorithm it is not absolutely necessary but convenient to start with a set of ortho*normal* polynomials. A set of orthogonal polynomials can be orthonormalized by dividing each polynomial by the square root of the integral of its square modulus, $\|p_l(x)\|_2$. If the monic orthogonal polynomials $p_l(x)$ are normalized by division by $\|p_l(x)\|_2$, the property of monicity gets lost. If $\tilde{p}_l(x)$ are the orthonormal polynomials, then

$$p_l(x) = \tilde{p}_l(x) \cdot \|p_l(x)\|_2. \tag{C.9}$$

So for the orthonormal polynomials $\tilde{p}_l(x)$, the recurrence relation (C.1) changes to

$$\tilde{p}_{l+1}(x) \cdot \|p_{l+1}(x)\|_2 = (x - a_l)\tilde{p}_l(x) \cdot \|p_l(x)\|_2 - b_l \tilde{p}_{l-1}(x) \cdot \|p_{l-1}(x)\|_2 \tag{C.10}$$

$$\Leftrightarrow$$

$$\tilde{p}_{l+1}(x) \cdot \frac{\|p_{l+1}(x)\|_2}{\|p_l(x)\|_2} = (x - a_l)\tilde{p}_l(x) - b_l \tilde{p}_{l-1}(x) \cdot \frac{\|p_{l-1}(x)\|_2}{\|p_l(x)\|_2} \tag{C.11}$$

Therefore, after the introduction of the new variable c_l, one can now formulate

$$x\tilde{p}_l(x) = c_{l+1}\tilde{p}_{l+1}(x) + a_l\tilde{p}_l(x) + c_l\tilde{p}_{l-1}(x), \text{ with } c_l = \sqrt{b_l} = \frac{\|p_l(x)\|_2}{\|p_{l-1}(x)\|_2}. \tag{C.12}$$

C The Golub–Welsch Algorithm

This can as well be written as the following matrix equation (cf. Gil et al. 2007, p. 142):

$$
x \begin{bmatrix} \tilde{p}_0(x) \\ \tilde{p}_1(x) \\ \tilde{p}_2(x) \\ \vdots \\ \tilde{p}_{n-1}(x) \end{bmatrix} = \begin{bmatrix} a_0 & c_1 & 0 & \cdots & 0 \\ c_1 & a_1 & c_2 & & \\ 0 & c_2 & a_2 & & \\ \vdots & & & & \vdots \\ 0 & \cdots & 0 & c_{n-1} & a_{n-1} \end{bmatrix} \cdot \begin{bmatrix} \tilde{p}_0(x) \\ \tilde{p}_1(x) \\ \tilde{p}_2(x) \\ \vdots \\ \tilde{p}_{n-1}(x) \end{bmatrix} + \begin{bmatrix} 0 \\ 0 \\ 0 \\ \vdots \\ c_n \tilde{p}_n(x) \end{bmatrix}. \tag{C.13}
$$

Looking at the short formulation of (C.13),

$$
x\tilde{p} = T \cdot \tilde{p} + c_n \tilde{p}_n e_{n-1}, \tag{C.14}
$$

it becomes obvious that (C.14) has close resemblance to the equation which the eigenvalues of the tridiagonal and symmetric matrix T would have to satisfy. In fact, the equation $\tilde{p}_n(\chi_l) = 0$ is equal to the statement that χ_l is an eigenvalue of T. So the sought roots χ_l of $\tilde{p}_n(x)$ are the eigenvalues of T and can be computed efficiently. The rank of T is at most equal to n, hence there do not exist more than the desired n eigenvalues χ_l which are in addition, as mentioned earlier, distinct from each other. Furthermore, because the geometric multiplicity of an eigenvalue never exceeds its algebraic multiplicity, one knows only one eigenvector is associated with each eigenvalue.

Since the polynomials $\tilde{p}_j(x)\tilde{p}_k(x)$, $l, k = 0, 1, \ldots, n - 1$ have degree smaller than $2n - 1$ the scalar products $\langle \tilde{p}_j(x), \tilde{p}_k(x) \rangle = \int_a^b \tilde{p}_j(x) \tilde{p}_k(x) w(x) dx$ can be evaluated by means of the Gauss quadrature. By virtue of the orthonormality condition, we have that

$$
\int_a^b \tilde{p}_j(x) \tilde{p}_k(x) w(x) \, dx = \sum_{l=1}^n \alpha_l \tilde{p}_j(\chi_l) \tilde{p}_k(\chi_l) = \begin{cases} 0, & \text{if } i \neq k \\ 1, & \text{if } j = k \end{cases}. \tag{C.15}
$$

In matrix notation this can be written as

$$
P^T W P = I, \tag{C.16}
$$

where

$$
W = \begin{bmatrix} \alpha_1 & 0 & 0 & \cdots & 0 \\ 0 & \alpha_2 & 0 & & \\ 0 & 0 & \alpha_3 & & \\ \vdots & & & \ddots & \\ 0 & \cdots & 0 & & \alpha_n \end{bmatrix} \text{ and } P = \begin{bmatrix} \tilde{p}_0(\chi_1) & \cdots & \tilde{p}_{n-1}(\chi_1) \\ \vdots & & \vdots \\ \tilde{p}_0(\chi_n) & \cdots & \tilde{p}_{n-1}(\chi_n) \end{bmatrix}. \tag{C.17}
$$

The n rows of the matrix P, from now on abbreviated as p_l, $l = 1, \ldots, n$, represent the eigenvectors of the Matrix T to the associated eigenvalues $\chi_1, \ldots, \chi_n$. From equation (C.16) it can be concluded that $W^{-1} = PP^T$, or in another notation,

$$\frac{1}{\alpha_l} = \sum_{j=0}^{n-1} (\tilde{p}_j(\chi_l))^2 = \|p_l\|_2^2, \quad l = 1, \ldots, n \tag{C.18}$$

The eigenvectors of T can also be calculated by solving the well-known equation (C.19) for v_l:

$$(T - I\chi_l) v_l = 0. \tag{C.19}$$

Each eigenvector is unique to its associated eigenvalue, up to a constant C. Therefore, $v_l = Cp_l$ or in long notation

$$(v_{1,l}, v_{2,l}, \ldots, v_{n,l})^T = C(\tilde{p}_0(\chi_l), \tilde{p}_1(\chi_l), \ldots, \tilde{p}_{n-1}(\chi_l))^T. \tag{C.20}$$

To calculate the constant C it is helpful to use the polynomial $\tilde{p}_0(\chi_l)$, since it is only a constant:

$$1 = \langle \tilde{p}_0(\chi_l), \tilde{p}_0(\chi_l) \rangle = \langle \tilde{p}_0(x), \tilde{p}_0(x) \rangle \tag{C.21}$$

$$= \tilde{p}_0(x)^2 \int_a^b w(x)\,dx = \tilde{p}_0(x)^2 \cdot \mu_0 \tag{C.22}$$

$$\Rightarrow \tilde{p}_0(x) = \frac{1}{\sqrt{\mu_0}}. \tag{C.23}$$

Equation (C.20) in connection with equation (C.23) yields that

$$\frac{v_{1,l}}{\tilde{p}_0(x)} = v_{1,l}\sqrt{\mu_0} = C \tag{C.24}$$

$$\Rightarrow v_l = v_{1,l}\sqrt{\mu_0}\, p_l \tag{C.25}$$

By the help of equation (C.18) one obtains finally:

$$\alpha_l = \mu_0 \cdot \frac{v_{1,l}^2}{\|v_l\|_2^2}. \tag{C.26}$$

Expressed in words this means that the weight α_l is equal to the squared first component of the normalized eigenvector v_l, multiplied by the integral of the weight function. As mentioned earlier, the Legendre and Hermite polynomials constructed according to (3.31) are monic. As to be seen in the previous exemplary calculations in Sects. 3.1.3 and 3.1.3, for odd exponents of x the integrals in the denominators are

C The Golub–Welsch Algorithm

zero. Therefore, the parameters a_l of matrix T in equation (C.13) are zero as well. With respect to the Legendre polynomials, the parameters which are needed for the implementation of the Golub–Welsch algorithm are

$$\mu_0 = 2, \quad a_l = 0, \quad c_l = \frac{l}{\sqrt{4l^2 - 1}}. \tag{C.27}$$

In case of the Hermite polynomials, the parameters are

$$\mu_0 = \sqrt{\pi}, \quad a_l = 0, \quad c_l = \sqrt{\frac{l}{2}}. \tag{C.28}$$

Appendix D
Simplified Multidimensional Moment Equations

To illustrate the properties of the restricted abscissae and weights with respect to the calculation of the equations

$$\left[M_1^{a_1} \odot M_2^{a_2} \odot \ldots \odot M_d^{a_d} \right] \cdot \alpha = \mathbb{E}\left[x_1^{a_1} x_2^{a_2} \ldots x_d^{a_d} \right], \quad \sum_{i=1}^{d} a_i \leq m, \tag{D.1}$$

the two different cases of odd and even sums of a_i have to be inspected. For the following remarks it shall be assumed that the rows M_i and associated exponents a_i are separated in two blocks of even and odd exponents. The rows M_i and exponents a_i are renamed according to these two partitions:

$$\begin{aligned} B_j^{u_j}, j &= 1, \ldots, k \text{ for even } a_i \\ C_j^{v_j}, j &= 1, \ldots, l \text{ for odd } a_i \\ k + l &= d. \end{aligned} \tag{D.2}$$

Case 1: The sum of the exponents a_i is odd.

Because all exponents u_j are even the sum of the exponents v_j must be odd and it follows that the number of exponents v_j is also odd.[1] Therefore, the vector

$$\left(C_1^{v_1} \odot C_2^{v_2} \odot \ldots \odot C_l^{v_l} \right) \tag{D.3}$$

---

[1] If $x_1 \ldots x_n$ are odd numbers and the sum of these numbers is odd, then n must be odd, too. This becomes clear by rewriting the sum to $\underbrace{x_1 - 1}_{\text{even}} + \underbrace{x_2 - 1}_{\text{even}} + \ldots + \underbrace{x_n - 1}_{\text{even}} + \underbrace{n}_{\text{odd}}$.

has symmetric absolute values but asymmetric signs. An example for $l = 3$:

$$C_1 = \begin{bmatrix} 3 & 4 & -4 & -3 \end{bmatrix}$$
$$C_2 = \begin{bmatrix} -1 & 2 & -2 & 1 \end{bmatrix} \quad \text{(D.4)}$$
$$C_3 = \begin{bmatrix} 5 & -1 & 1 & -5 \end{bmatrix}.$$

The exponents v_j are odd and therefore the signs within the vectors $C_j^{v_j}$ are equal to those within C_j. Consequently, because the signs within the resulting vector

$$\left(C_1^{v_1} \odot C_2^{v_2} \odot C_3^{v_3} \right) \quad \text{(D.5)}$$

are in the focus of investigation, there is no need to consider the exponents any further. It is sufficient to observe the product

$$(C_1 \odot C_2 \odot C_3) = \begin{bmatrix} 3 \cdot -1 \cdot 5 & 4 \cdot 2 \cdot -1 & -4 \cdot -2 \cdot 1 & -3 \cdot 1 \cdot -5 \end{bmatrix}$$
$$= \begin{bmatrix} -15 & -8 & 8 & 15 \end{bmatrix}. \quad \text{(D.6)}$$

This example clarifies that, if l is odd, the product (D.3) necessarily yields a row vector with symmetric absolute values and asymmetric signs (observe the "columns" of (D.4)). The reason for this fact lies in the restricted structure of the vectors M_j.

The product

$$\left(B_1^{u_1} \odot B_2^{u_2} \odot \ldots \odot B_k^{u_k} \right) \quad \text{(D.7)}$$

obviously yields a row vector with symmetric and positive values, because all exponents u_i are even. As a consequence, also the vector

$$\left(M_1^{a_1} \odot M_2^{a_2} \odot \ldots \odot M_d^{a_d} \right)$$
$$= \left(B_1^{u_1} \odot B_2^{u_2} \odot \ldots \odot B_k^{u_k} \right) \odot \left(C_1^{v_1} \odot C_2^{v_2} \odot \ldots \odot C_l^{v_l} \right) \quad \text{(D.8)}$$

has symmetric absolute values but asymmetric signs. Concluding, one obtains the result that

$$\left(M_1^{a_1} \odot M_2^{a_2} \odot \ldots \odot M_d^{a_d} \right) \cdot \alpha = 0 \quad \text{(D.9)}$$

for all choices of restricted abscissae χ_l and associated restricted weights α_l.

Case 2: The sum of the exponents a_i is even.

By arguments which are analogous to the first case, the second case can be investigated. All exponents u_j are even and so the sum of the exponents v_j must be even. It follows that the number of exponents v_j is also even. Therefore, the

D Simplified Multidimensional Moment Equations

resulting vector

$$\left(C_{1_v}^{v_1} \odot C_{2_v}^{v_2} \odot \ldots \odot C_{l_v}^{v_l}\right) \tag{D.10}$$

has symmetric absolute values and symmetric signs. According to this, also the vector

$$\begin{aligned}&\left(M_1^{a_1} \odot M_2^{a_2} \odot \ldots \odot M_d^{a_d}\right) \\ &= \left(B_1^{u_1} \odot B_2^{u_2} \odot \ldots \odot B_k^{u_k}\right) \odot \left(C_1^{v_1} \odot C_2^{v_2} \odot \ldots \odot C_l^{v_l}\right)\end{aligned} \tag{D.11}$$

has symmetric absolute values and symmetric signs. In conclusion, one obtains the result that the outcome of

$$\left(M_1^{a_1} \odot M_2^{a_2} \odot \ldots \odot M_d^{a_d}\right) \cdot \boldsymbol{\alpha} \tag{D.12}$$

depends on the choice of restricted abscissae χ_l and the associated restricted weights α_l.

Bibliography

Abramowitz, M., & Stegun, I. A. (1972). *Handbook of mathematical functions with formulas, graphs and mathematical tables* (10th ed.). New York: Dover Publications.

Andrews, A. (1968). A square root formulation of the Kalman covariance equations. *AIAA Journal, 6*(6), 1165–1166.

Antia, H. M. (1995). *Numerical methods for scientists and engineers* (2nd ed.). New Delhi: Tata McGraw-Hill Publishing. [rev.] edition.

Arasaratnam, I., & Haykin, S. (2009). Cubature Kalman filters. *IEEE Transactions on Automatic Control, 54*(6), 1254–1269.

Arasaratnam, I., Haykin, S., & Hurd, T. R. (2010). Cubature Kalman filtering for continuous-discrete systems: theory and simulations. *IEEE Transactions on Signal Processing, 58*(10), 4977–4993.

Arnold, L. (1974). *Stochastic differential equations: theory and applications*. New York: Wiley.

Arulampalam, M. S., Maskell, S., Gordon, N., & Clapp, T. (2002). A tutorial on particle filters for online nonlinear/non-Gaussian Bayesian tracking. *IEEE Transactions on Signal Processing, 50*(2), 174–188.

Bayes, T., & Price, R. (1763). An essay towards solving a problem in the doctrine of chances. By the late Rev. Mr. Bayes, F. R. S. communicated by Mr. Price, in a letter to John Canton, A. M. F. R. S. *Philosophical Transactions of the Royal Society of London, 53*(0), 370–418.

Beckers, M., & Haegemans, A. (1991). The construction of three-dimensional invariant cubature formulae. *Journal of Computational and Applied Mathematics, 35*(1–3), 109–118.

Bellantoni, J. F., & Dodge, K. W. (1967). A square root formulation of the Kalman-Schmidt filter. *AIAA Journal, 5*(7), 1309–1314.

Bergstrom, A. R. (1966). Nonrecursive models as discrete approximations to systems of stochastic differential equations. *Econometrica, 34*(1), 173.

Boussaïd, I., Lepagnot, J., & Siarry, P. (2013). A survey on optimization metaheuristics. *Information Sciences, 237*, 82–117.

Box, G. E., Jenkins, G. M., & Reinsel, G. C. (2008). *Time series analysis: forecasting and control* (4th ed.). Wiley series in probability and statistics. Hoboken, NJ: John Wiley and Sons.

Bronštejn, I. N., Musiol, G., & Mühlig, H. (2005). *Taschenbuch der Mathematik*. Deutsch, Frankfurt am Main, 6., vollst. überarb. und erg. aufl edition.

Brooks, C. (1998). Chaos in foreign exchange markets: a sceptical view. *Computational Economics, 11*(3), 265–281.

Burkhardt, J. (2014). *Slow exponential growth for Clenshaw Curtis sparse grids*. Tech. Rep., Virginia Tech.

Carlin, B. P., Polson, N. G., & Stoffer, D. S. (1992). A Monte Carlo approach to nonnormal and nonlinear state-space modeling. *Journal of the American Statistical Association, 87*(418), 493.

Clenshaw, C. W., & Curtis, A. R. (1960). A method for numerical integration on an automatic computer. *Numerische Mathematik, 2*(1), 197–205.

Cools, R. (1997). Constructing cubature formulae: the science behind the art. *Acta Numerica, 6*, 1.

Cools, R. (1999). Monomial cubature rules since "Stroud": a compilation — part 2. *Journal of Computational and Applied Mathematics, 112*(1–2), 21–27.

Cools, R. (2003). An encyclopaedia of cubature formulas. *Journal of Complexity, 19*(3), 445–453.

Cools, R., & Rabinowitz, P. (1993). Monomial cubature rules since "Stroud": a compilation. *Journal of Computational and Applied Mathematics, 48*(3), 309–326.

Cools, R., Novak, E., & Ritter, K. (1998). Smolyak's construction of cubature formulas of arbitrary trigonometric degree. *Computing, 62*(2), 147–162.

Ehrich, S. (1995). Asymptotic properties of Stieltjes polynomials and Gauss-Kronrod quadrature formulas. *Journal of Approximation Theory, 82*(2), 287–303.

Elhay, S., & Kautsky, J. (1992). Generalized Kronrod Patterson type imbedded quadratures. *Applications of Mathematics, 37*(2), 81–103.

Engels, H. (1980). *Numerical quadrature and cubature.* Computational mathematics and applications. London and New York: Academic Press.

Evensen, G. (1994). Sequential data assimilation with a nonlinear quasi-geostrophic model using Monte Carlo methods to forecast error statistics. *Journal of Geophysical Research, 99*(C5), 10143.

Evensen, G. (1997). Advanced data assimilation for strongly nonlinear dynamics. *Monthly Weather Review, 125*(6), 1342–1354.

Evensen, G. (2003). The ensemble Kalman filter: theoretical formulation and practical implementation. *Ocean Dynamics, 53*(4), 343–367.

Faggini, M., & Parziale, A. (2012). The failure of economic theory. Lessons from Chaos theory. *Modern Economy, 03*(01), 1–10.

Fernández-Villaverde, J. (2010). The econometrics of DSGE models. *SERIEs, 1*(1–2), 3–49.

Florescu, I. (2014). *Probability and stochastic processes.* New York: Wiley.

Gautschi, W. (1968). Construction of Gauss-Christoffel quadrature formulas. *Mathematics of Computation, 22*(102), 251.

Gautschi, W., & Notaris, S. E. (1996). Stieltjes polynomials and related quadrature formulae for a class of weight functions. *Mathematics of Computation, 65*(215), 1257–1269.

Gelb, A. (1974). *Applied optimal estimation.* Cambridge, MA: MIT Press.

Genz, A., & Keister, B. D. (1996). Fully symmetric interpolatory rules for multiple integrals over infinite regions with Gaussian weight. *Journal of Computational and Applied Mathematics, 71*(2), 299–309.

Gil, A., Segura, J., & Temme, N. M. (2007). *Numerical methods for special functions.* Society for Industrial and Applied Mathematics (SIAM, 3600 Market Street, Floor 6, Philadelphia, PA 19104). Philadelphia, PA: SIAM.

Ginzburg, V. L., & Landau, L. D. (1950). On the theory of superconductivity. *Soviet Physics, JETP, 20*, 1064–1082.

Golub, G. H., & Welsch, J. H. (1969). Calculation of Gauss quadrature rules. *Mathematics of Computation, 23*(106), 221.

Gradshtein, I. S., Ryzhik, I. M., Jeffrey, A., & Zwillinger, D. (op. 2007). *Table of integrals, series, and products* (7th ed.). Amsterdam and Boston and Paris [et al.]: Elsevier and Academic Press.

Gubner, J. A. (2009). *Gaussian quadrature and the eigenvalue problem.* Tech. Rep., University of Wisconsin.

Guégan, D., & Mercier, L. (2006). Prediction in chaotic time series: methods and comparisons with an application to financial intra-day data. *The European Journal of Finance, 11*(2), 137–150.

Haegemans, A., & Piessens, R. (1976). Construction of cubature formulas of degree eleven for symmetric planar regions, using orthogonal polynomials. *Numerische Mathematik, 25*(2), 139–148.

Hallegatte, S., Ghil, M., Dumas, P., & Hourcade, J.-C. (2008). Business cycles, bifurcations and chaos in a neo-classical model with investment dynamics. *Journal of Economic Behavior & Organization, 67*(1), 57–77.

Hammer, P. C., & Stroud, A. H. (1958). Numerical evaluation of multiple integrals. II. *Mathematics of Computation, 12*(64), 272.

Hammer, P. C., & Wymore, A. W. (1957). Numerical evaluation of multiple integrals I. *Mathematical Tables and Other Aids to Computation, 11*(58), 59.

Heiss, F., & Winschel, V. (2008). Likelihood approximation by numerical integration on sparse grids. *Journal of Econometrics, 144*(1), 62–80.

Hinrichs, A., & Novak, E. (2007). Cubature formulas for symmetric measures in higher dimensions with few points. *Mathematics of Computation, 76*(259), 1357–1373.

Ito, K., & Xiong, K. (2000). Gaussian filters for nonlinear filtering problems. *IEEE Transactions on Automatic Control, 45*(5), 910–927.

Jazwinski, A. H. (2007). *Stochastic processes and filtering theory* (dover ed.). Dover books on engineering. Mineola, NY: Dover Publications.

Jetter, K. (2006). *Topics in multivariate approximation and interpolation*, volume 12 of *Studies in computational mathematics* (1st ed.). Amsterdam and Boston: Elsevier.

Jia, B., Xin, M., & Cheng, Y. (2013). High-degree cubature Kalman filter. *Automatica, 49*(2), 510–518.

Judd, K. L. (1998). *Numerical methods in economics*. Cambridge, MA: MIT Press.

Julier, S. J., Uhlmann, J., & Durrant-Whyte, H., editors (1995). *A new approach for filtering nonlinear systems*. Piscataway and N.J. Distributed through the IEEE Service Center.

Kalman, R. E. (1960). A new approach to linear filtering and prediction problems. *Journal of Basic Engineering, 82*(1), 35.

Kalman, R. E. (1963). Mathematical description of linear dynamical systems. *Journal of the Society for Industrial and Applied Mathematics Series A Control, 1*(2), 152–192.

Kitagawa, G. (1987). Non-Gaussian state-space modeling of nonstationary time series: rejoinder. *Journal of the American Statistical Association, 82*(400), 1060.

Klebaner, F. C. (2005). *Introduction to stochastic calculus with applications* (2nd ed.). London and Singapore: Imperial College Press and Distributed by World Scientific Publication.

Kloeden, P. E., & Platen, E. (1992). *Numerical solution of stochastic differential equations*, volume 23 of *Applications of mathematics*. Berlin and New York: Springer-Verlag.

Krommer, A. R., & Ueberhuber, C. W. (1998). *Computational integration*. Philadelphia: Society for Industrial and Applied Mathematics.

Kronrod, A. S. (1965). *Nodes and weights of quadrature formulas: Sixteen-place tables*. New York: Consultants Bureau.

Kuperberg, G. (2006). Numerical cubature using error-correcting codes. *SIAM Journal on Numerical Analysis, 44*(3), 897–907.

Laurie, D. P. (1997). Calculation of Gauss-Kronrod quadrature rules. *Mathematics of Computation, 66*(219), 1133–1146.

Lawson, C. L., & Hanson, R. J. (1995). *Solving least squares problems*, volume 15 of *Classics in applied mathematics*. Philadelphia, PA: SIAM. Unabridged, rev. republ. (of the 1974 ed.) edition.

Liptser, R. S., & Shiryaev, A. N. (2001). *Statistics of random processes: II. applications*, volume 6 of *Stochastic modelling and applied probability*. Berlin Heidelberg: Springer. Second, rev. and expanded edition.

Lorenz, E. N. (1963). Deterministic nonperiodic flow. *Journal of the Atmospheric Sciences, 20*(2), 130–141.

Lorenz, H.-W. (1987). Strange attractors in a multisector business cycle model. *Journal of Economic Behavior & Organization, 8*(3), 397–411.

Lyness, J. L., & Cools, R. (1994). A Survey of numerical Quadrature over Triangles. In *Proceedings of Symposia in Applied Mathematics* (Vol. 48), Argonne.

Majorana, C., Odorizzi, S., & Vitaliani, R. (1982). Shortened quadrature rules for finite elements. *Advances in Engineering Software (1978), 4*(2), 52–57.

Mason, J. C., & Handscomb, D. C. (2003). *Chebyshev polynomials*. Boca Raton, FL: Chapman & Hall/CRC.
Matilla-García, M., & Marín, M. R. (2010). A new test for chaos and determinism based on symbolic dynamics. *Journal of Economic Behavior & Organization, 76*(3), 600–614.
Mehrotra, S., & Papp, D. (2012). Generating nested Quadrature Formulas for general Weight Functions with known moments. arXiv preprint arXiv:1203.1554.
Möller, H. M. (1976). Kubaturformeln mit minimaler Knotenzahl. *Numerische Mathematik, 25*(2), 185–200.
Möller, H. M. (1979). Lower bounds for the number of nodes in cubature formulae. In G. Hämmerlin (Ed.), *Numerische integration*, volume 45 of *ISNM* (pp. 221–230). Birkhäuser: Basel.
Monegato, G. (1978). Positivity of the weights of extended Gauss-Legendre quadrature rules. *Mathematics of Computation, 32*(141), 243.
Monegato, G. (2001). An overview of the computational aspects of Kronrod quadrature rules. *Numerical Algorithms, 26*(2), 173–196.
Novak, E., & Ritter, K. (1999). Simple cubature formulas with high polynomial exactness. *Constructive Approximation, 15*(4), 499–522.
Olver, Frank W. J. (2010). *NIST handbook of mathematical functions: companion to the digital library of mathematical functions*. Cambridge: Cambridge University Press.
Patterson, T. N. (1968). The optimum addition of points to quadrature formulae. *Mathematics of Computation, 22*(104), 847.
Petras, K. (2003). Smolyak cubature of given polynomial degree with few nodes for increasing dimension. *Numerische Mathematik, 93*(4), 729–753.
Phillips, G. M. (1967). Numerical integration over an N-dimensional rectangular region. *The Computer Journal, 10*(3), 297–299.
Potter, J., & Stern, R. (1963). Statistical filtering of space navigation measurements. In *Guidance and Control Conference*.
Rabinowitz, P. (1980). The exact degree of precision of generalized Gauss-Kronrod integration rules. *Mathematics of Computation, 35*(152), 1275.
Savona, R., Soumare, M., & Andersen, J. V. (2015). Financial symmetry and moods in the market. *PloS one, 10*(4), e0118224.
Schmidt, E. (1908). Über die Auflösung Linearer Gleichungen mit Unendlich Vielen Unbekannten. *Rendiconti del Circolo Matematico di Palermo, 25*, 53–77.
Schmidt, S. F. (1966). *Application of state-space methods to navigation problems*, volume 3 of *Advances in control systems* (pp. 293–340). New York: Academic Press.
Schweppe, F. (1965). Evaluation of likelihood functions for Gaussian signals. *IEEE Transactions on Information Theory, 11*(1), 61–70.
Shafarevich, I. R., & Remizov, A. O. (2013). *Linear algebra and geometry*. Berlin and New York: Springer.
Shaffer, S. (1991). Structural shifts and the volatility of chaotic markets. *Journal of Economic Behavior & Organization, 15*(2), 201–214.
Sickel, W., & Ullrich, T. (2006). *Smolyak's algorithm, sampling on sparse grids and function spaces of dominating mixed smoothness*. Jenaer Schriften zur Mathematik und Informatik (Vol. 2006). Jena: Friedrich-Schiller-University.
Singer, H. (1990). *Parameterschätzung in zeitkontinuierlichen dynamischen Systemen (Parameter estimation in continuous time dynamical systems) (Ph.D. Thesis)*, Hartung-Gorre-Verlag, Konstanz.
Singer, H. (1999). *Finanzmarktökonometrie: Zeitstetige Systeme und ihre Anwendung in Ökonometrie und empirischer Kapitalmarktforschung*, volume 171 of *Wirtschaftswissenschaftliche Beiträge*. Heidelberg: Physica.
Singer, H. (2006a). Continuous-discrete unscented Kalman filtering. *Diskussionsbeiträge Fakultät für Wirtschaftswissenschaft FernUniversität in Hagen* (384).
Singer, H. (2006b). Moment equations and Hermite expansion for nonlinear stochastic differential equations with application to stock price models. *Computational Statistics, 21*(3–4), 385–397.

Singer, H. (2008). Generalized Gauss–Hermite filtering. *AStA Advances in Statistical Analysis, 92*(2), 179–195.

Singer, H. (2011). Continuous-discrete state-space modeling of panel data with nonlinear filter algorithms. *AStA Advances in Statistical Analysis, 95*(4), 375–413.

Singer, H. (2015). Conditional Gauss–Hermite filtering with application to volatility estimation. *IEEE Transactions on Automatic Control, 60*(9), 2476–2481.

Smolyak, S. A. (1963). Quadrature and interpolation formulas for tensor products of certain classes of functions. *Soviet Mathematics Doklady, 4*, 240–243.

Stenger, F. (1971). Tabulation of certain fully symmetric numerical integration formulas of degree 7, 9 and 11. *Mathematics of Computation, 25*(116), 935+s58–s125.

Stroud, A. H. (1966). Some fifth degree integration formulas for symmetric regions. *Mathematics of Computation, 20*(93), 90.

Stroud, A. H. (1967a). Some fifth degree integration formulas for symmetric regions II. *Numerische Mathematik, 9*(5), 460–468.

Stroud, A. H. (1967b). Some seventh degree integration formulas for symmetric regions. *SIAM Journal on Numerical Analysis, 4*(1), 37–44.

Stroud, A. H. (1971). *Approximate calculation of multiple integrals*. Prentice-Hall series in automatic computation. Englewood Cliffs, NJ: Prentice Hall.

Stroud, A. H., & Secrest, D. (1963). Approximate integration formulas for certain spherically symmetric regions. *Mathematics of Computation, 17*(82), 105.

Szegö, G. (1935). Über gewisse orthogonale Polynome, die zu einer oszillierenden Belegungsfunktion gehören. *Mathematische Annalen, 110*(1), 501–513.

Tang, K.-T. (2007). *Mathematical methods for engineers and scientists 2: vector analysis, ordinary differential equations and Laplace transforms*. Berlin, Heidelberg: Springer-Verlag.

Tanizaki, H. (2010). *Nonlinear filters: Estimation and applications; with 45 tables* (2nd ed.). Berlin [u.a.]: Springer.

Tanizaki, H. (op. 2004). *Computational methods in statistics and econometrics*. New York and London: Marcel Dekker and Taylor & Francis.

Trefethen, L. N. (2008). Is Gauss quadrature better than Clenshaw–Curtis? *SIAM Review, 50*(1), 67–87.

Tyler, G. W. (1953). Numerical integration of functions of several variables. *Canadian Journal of Mathematics, 5*(0), 393–412.

van der Merwe, R., & Wan, E. (Eds.). (2001). *IEEE International Conference on Acoustics, Speech, and Signal Processing* (Vol. 2001), 7–11 May 2001, Salt Palace Convention Center, Salt Lake City, Utah, USA/sponsored by the Institute of Electrical and Electronics Engineers, Signal Processing Society. In *Proceedings* (Vol. 1). Piscataway, NJ: IEEE Operations Center.

van Loan, C. (1978). Computing integrals involving the matrix exponential. *IEEE Transactions on Automatic Control, 23*(3), 395–404.

Vialar, T. (2015). *Handbook of mathematics*. Paris: Books on demand.

Victoir, N. (2004). Asymmetric cubature formulae with few points in high dimension for symmetric measures. *SIAM Journal on Numerical Analysis, 42*(1), 209–227.

Vlad, S., Pascu, P., & Morariu, N. (2010). Chaos models in economics. *Journal of Computing, 2*(1), 79–83.

Waldvogel, J. (2006). Fast construction of the Fejér and Clenshaw–Curtis quadrature rules. *BIT Numerical Mathematics, 46*(1), 195–202.

Wasilkowski, G. W., & Wozniakowski, H. (1995). Explicit cost bounds of algorithms for multivariate tensor product problems. *Journal of Complexity, 11*(1), 1–56.

Wilf, H. S. (1978, c1962). *Mathematics for the physical sciences*. New York: Dover Publications.

Winschel, V., & Krätzig, M. (2010). Solving, estimating, and selecting nonlinear dynamic models without the curse of dimensionality. *Econometrica, 78*(2), 803–821.

Wright, M. H. (2005). The interior-point revolution in optimization: history, recent developments, and lasting consequences. *Bulletin of the American Mathematical Society, 42*(01), 39–57.

Zwillinger, D. (1998). *Handbook of differential equations* (volume Bd. 1). New York: Academic Press.

Index

AE rules, 96, 135

Bayes filter, 6
Bayesian measurement update, 7
Bayesian parameter estimation, 22
Bayesian time update, 7

Centrally symmetric integral, 68
Chaotic system, 126
Chapman-Kolmogorov Equation, 7
Clenshaw-Curtis quadrature, 62
Complete d-dimensional polynomials, 68
Compound rules, 88
Conditional filtering, 31
Converged mean absolute deviation, 110
Cubature, 2, 65
Cubature Kalman filter, 77

Degree of polynomial exactness, 2
Degrees of multidimensional polynomials, 68
Delayed Gauss-Kronrod-Patterson rules, 86
Deterministic numerical integration, 2, 48, 65
Discrete Bayes filter, 7
Drift function, 6
Dynamical system, 5

Exact discrete model, 120

Gauss quadrature, 53
Gauss-Hermite quadrature, 56

Gauss-Kronrod quadrature, 56
Gauss-Legendre quadrature, 55
Gauss-Patterson quadrature, 59
Gaussian assumption, 13
Gaussian filters, 14
General discrete state-space model, 6
Genz-Keister rules, 62
Ginzburg-Landau model, 44, 130
Golub-Welsch algorithm, 55, 143
Gram-Schmidt orthogonalization algorithm, 54, 56

Hermite polynomials, 56

Kalman filter, 1, 10
Kalman filter for linear state-space models, 12
Kalman gain, 10

Lagrange interpolation, 48
Lagrange polynomials, 49
Legendre polynomials, 55
Lorenz model, 126
Lower bound of Möller, 67

Maximum Likelihood estimation, 21
Mean absolute deviation, 110, 115
Measurement equation, 6
Method of van Loan, 119
Missing Data, 43
Multidimensional deterministic numerical integration, 65

Multidimensional moment equations, 71
Multidimensional monomials, 68
Multidimensional polynomials, 68

Nonlinear Kalman filter, 13, 110

One-dimensional deterministic numerical integration, 48
Orthogonal polynomials, 54–56

Parameter estimation, 21
Polynomial exactness, 2
Product cubature rules, 69

Quadrature, 2

Riemann sums, 7

Six-dimensional coordinated turn model, 117
Smolyak cubature, 80

Smolyak-AE rules, 107
Stability Factor, 66
Stability of cubature rules, 2, 98
Stabilized(1) rules, 4, 99, 135
Stabilized(2) rules, 4, 102, 136
State equation, 6
State vector, 2, 5
State-space model, 5

Tensor product, 69
Theorem on normal Correlation, 10

Univariate non-stationary growth model, 112
Unscented conditional Kalman filter algorithm, 38
Unscented Kalman filter, 3, 17
Unscented transform, 3, 18, 80

Variation of parameters, 118
Vector space, 53, 68

Weight function, 47